Diagnostic Endoscopy

Series in Medical Physics and Biomedical Engineering

Series Editors: John G Webster, E Russell Ritenour, Slavik Tabakov, Kwan-Hoong Ng, and Alisa Walz-Flannigan

Series in Medical Physics and Biomedical Engineering

Diagnostic Endoscopy

Edited by
Haishan Zeng
BC Cancer Research Centre
University of British Columbia
Vancouver, British Columbia, Canada

CRC Press
Taylor & Francis Group
Boca Raton London New York

CRC Press is an imprint of the
Taylor & Francis Group, an **informa** business

CRC Press
Taylor & Francis Group
6000 Broken Sound Parkway NW, Suite 300
Boca Raton, FL 33487-2742

First issued in paperback 2019

© 2014 by Taylor & Francis Group, LLC
CRC Press is an imprint of Taylor & Francis Group, an Informa business

No claim to original U.S. Government works

ISBN-13: 978-1-4200-8346-0 (hbk)
ISBN-13: 978-0-367-37907-0 (pbk)

Visit the Taylor & Francis Web site at
http://www.taylorandfrancis.com

and the CRC Press Web site at
http://www.crcpress.com

Dedication

To my parents for their love and encouragement.

To my wife, Hui, my children, Demi, Aaron, and Felicity, for their love and support.

Contents

SECTION I Basics of Endoscopy and Light-Tissue Interactions

SECTION II Endoscopic Field Imaging Modalities

SECTION III Endoscopic Point Spectroscopy

SECTION IV   Endoscopic "Point" Microscopy

SECTION V   Clinical Applications

Preface

The invention of the flexible fiber optic endoscope 55 years ago in 1958 and the subsequent development in the 1960s revolutionized medical diagnosis and treatment of internal organs. Since then there have been many technological advancements, such as the introduction of video chip endoscopy. However, its imaging mechanism of broad-band white light reflectance has undergone almost no changes in the first three decades after its invention. The invention of fluorescence endoscopy in early 1990s represented the first significant addition of new imaging mechanisms to endoscopy. This is a result of improved understanding of tissue optical properties. During the last 30 years, there was a surge on the study of light-tissue interactions and tissue optical properties as part of a new subject—biomedical optics and biophotonics. These researches and related technology developments have resulted in new optical spectroscopy and imaging modalities (mechanisms) that can be used with endoscope to increase diagnostic yield and facilitate better therapeutic interventions. They include both new field imaging modalities for improving diagnostic sensitivity and powerful point analysis spectroscopy and "point" microscopy imaging techniques for improving diagnostic specificity, thanks to their capabilities of deriving tissue biochemistry and micro-morphology information. This book is intended to present a systematic coverage of these new spectroscopy and imaging modalities.

The material here has been developed partially from a technical course "Diagnostic Endoscopy" I taught for SPIE, the International Society for Optics and Photonics, and a graduate course, "Biomedical Optics," I taught at the Department of Physics and Astronomy, University of British Columbia, plus six chapters written by invited contributors. Its intent is to serve as an introductory guide for students, engineers, and researchers who are interested in developing optical diagnostic applications through endoscopes. Those who are thinking of how to adapt their own optical technologies for endoscopic applications may find this book particularly valuable.

I wish to thank my mentors and long-term collaborators at the British Columbia Cancer Agency and the University of British Columbia, especially, Dr. Branko Palcic, Dr. David I. McLean, Dr. Calum MacAulay, Dr. Stephen Lam, and Dr. Harvey Lui. I benefited enormously from the more than 20 years of intellectual interactions with them. I would also like to thank all the students, fellows, researchers, and engineers who worked with me over the years. Without your hard work, many of the cited work in this book would not have been accomplished. Finally, I am grateful to all the invited contributors and CRC Press staff involved in the book's editing process.

Haishan Zeng, PhD
Vancouver, British Columbia
Canada
June 2013

The Editor

Haishan Zeng is a distinguished scientist with the Imaging Unit, Integrative Oncology Department, British Columbia Cancer Agency Research Centre, and a professor of dermatology, pathology, and physics at the University of British Columbia, Vancouver, Canada. Dr. Zeng received a B.Sc. degree in electronic physics from Peking University, Beijing, China, a M.Sc. degree in electronic physics and devices from the Chinese Academy of Sciences, and a Ph.D. degree in biophysics from the University of British Columbia. Dr. Zeng has been associated with the British Columbia Cancer Agency and the University of British Columbia for more than 24 years, where his research has been focused on the optical properties of biological tissues and light-tissue interaction, as well as their applications in medical diagnosis and therapy. He has published over 105 refereed journal papers and eight book chapters, and has edited eight conference proceedings. He has 21 awarded patents and a dozen patent applications related to optical diagnosis and therapy. Three medical devices derived from these patents, have passed regulatory approvals, and are currently in clinical uses around the world. The latest device, Verisante Aura™ using Raman spectroscopy for noninvasive skin cancer detection, was awarded the Prism Award in the Life Sciences and Biophotonics category in February 2013 by SPIE, the International Society for Optics and Photonics.

Contributors

Mu Chiao
University of British Columbia
Vancouver, British Columbia, Canada

Michele Follen
Texas Tech University
El Paso, Texas

Pierre Lane
British Columbia Cancer Agency
    Research Centre
Vancouver, British Columbia, Canada

Xingde Li
Johns Hopkins University
Baltimore, Maryland

Wenxuan Liang
Johns Hopkins University
Baltimore, Maryland

Calum MacAulay
British Columbia Cancer Agency
    Research Centre
Vancouver, British Columbia, Canada

Hadi Mansoor
University of British Columbia,
Vancouver, British Columbia, Canada

Jessica Mavadia
Johns Hopkins University
Baltimore, Maryland

Annette McWilliams
British Columbia Cancer Agency
    Research Centre
Vancouver, British Columbia, Canada

Naoki Muguruma
St. Michael's Hospital
Toronto, Ontario, Canada

Kartikeya Murari
Johns Hopkins University
Baltimore, Maryland

Beau Standish
Ryerson University
Toronto, Canada

Victor Young
Ryerson University and Sunnybrook
    Health Science Centre
Toronto, Ontario, Canada

Haishan Zeng
Imaging Unit, Integrative Oncology
    Department
British Columbia Cancer Agency
    Research Centre
Vancouver, British Columbia, Canada

Section I

Basics of Endoscopy and Light-Tissue Interactions

1 Introduction

Haishan Zeng
British Columbia Cancer Agency Research Centre,
Vancouver, British Columbia, Canada

CONTENTS

1.1 HISTORY OF ENDOSCOPY

The word *endoscopy* was adopted from Greek, meaning "to examine within" (Benedict 1951; Berci 1976). The endoscope is an instrument for examining the interior of a hollow viscus. In modern medicine it also facilitates therapeutic interventions. The earliest concept of the endoscope can be traced back to the 1800s according to Berci (1976). In 1806, Bozzini tried to developed a method for examination of deeply seated organs. He used a wax candle as the light source, a shaped tin tub, and a reflector mirror. In 1868, Bevan extracted foreign objects and saw the esophagus strictures and tumors. In 1870, Kussmaul demonstrated the first endoscopy (esophagogastroscopy) examination employing a volunteer professional sword swallower as the "patient." This "open-tube endoscope" had been further developed and refined for many years. For example, two tubes were combined together for easier introduction.

The first key step in the evolution of endoscopy was the invention of an optical means (telescopes) to transmit the image from the deeply located organ to the outside. In 1879, Nitze teaming with Beneche (an optician) and Leiter (an instrument maker), successfully made the first cystoscope. Many telescope lens pairs mounted inside a rigid tube were used to relay the image. The imaging quality of the relayed images was greatly improved with the invention of Hopkin's rod-lens system replacing standard lenses. Modern rigid endoscopes are still using the Hopkin's rod-lenses (Berci 1976).

In 1936, Schindler teamed with Wolf, an optical physicist and manufacturer, developed a semi-flexible endoscope that still uses the telescope principles for image relay. The scope contained 48 lenses. The flexible portion consisted of steel spiral with lenses that were kept in place by a special spring covered by rubber tubes

3

outside. For more details on the early history of endoscopy development, there are a couple of good treatises (e.g., Benedict 1951; Berci 1976; Cotton and Williams 2003; Nelson 1970).

The invention of the flexible fiber optic endoscope by Hirchowitz et al. (1958) and the subsequent development in the 1960s revolutionized medical diagnosis and treatment of internal organs. Both the illumination light and the images are transmitted by flexible glass fiber bundles: the former through an incoherent fiber bundle (fibers arranged randomly), the latter through a coherent fiber bundle (fibers in both ends of the bundle keep the same ordered arrangement). Biopsy/instrument channels are also conveniently added. The endoscope is flexible and makes possible being able to reach various locations not accessible by previous telescope-based endoscopes. At this point, it is worth to mention the prior research and development activities that formed the basis of the invention of the flexible fiber optic endoscope. The idea of transmitting light through glass fibers was first patented in 1928 by Baird (see Nelson 1970), but its practical uses were delayed until coating a clad layer on the core fiber could be used to improve the light transmission (van Heel 1954) and a practical method of fabricating fibers was discovered (Hopkins and Kapany 1954).

The next important technical advancement occurred in the late 1980s, with the introduction of video-chip technology. The video endoscope has the CCD video chip installed at the distal tip of the endoscope, eliminated the imaging fiber bundle and resulting in much better imaging quality. A fiber optic endoscopy has 20,000 to 40,000 individual fibers in the imaging bundle, while the CCD sensor in a high definition format video endoscope has $1024 \times 768 = 786,432$ pixels. This and other advantages lead to the eventual replacement of fiber optic endoscopes by video endoscopes.

Other important developments include the invention of fluorescence endoscopy in the 1990s and capsule endoscopy in the 2000s. During the last 30 years, there was a surge in the study of light-tissue interactions and tissue optical properties. These researches and related technology developments have resulted in new optical spectroscopy and imaging modalities that can be used with the endoscope to increase diagnostic yield and facilitate better therapeutic interventions. It is the aim of this book to cover these developments.

1.2   DEFINITION OF DIAGNOSTIC PERFORMANCE

The performance of a diagnostic method (test) is usually measured in terms of *sensitivity*, *specificity*, positive predictive value (PPV), negative predictive value (NPV), and receiver operating characteristic (ROC) curves and their respective areas under the curve (AUC). Table 1.1 illustrates the definitions for some of the terms. The purpose of a diagnostic test is to separate disease cases from nondisease cases, which are determined by the gold standard (e.g., histopathology results of biopsied samples). But the diagnostic performance is usually not perfect. Therefore, a certain number (*a*) of disease cases are diagnosed as positive (true positive), while the rest (*c*) are diagnosed as negative (false negative). Similarly, a certain number (*d*) of nondisease cases are diagnosed as negative (true negative), while the rest (*b*) are diagnosed as positive (false positive). The diagnostic sensitivity is defined as

TABLE 1.1

Definition of Diagnostic Performance

	Disease	Nondisease	
Diagnosis positive	a True positive	b False positive	Positive Predictive Value (PPV) $= a/(a + b)$
Diagnosis negative	c False negative	d True negative	Negative Predictive Value (NPV) $= d/(c + d))$
	Sensitivity $= a/(a + c)$	Specificity $= d/(b + d)$	$a + b + c + d = N$ (total cases)

$$\text{sensitivity} = \text{true positives/total diseases} = a/(a+c)$$

Sensitivity is the ability to correctly identify those with the disease (true positive rate). It is different from the positive predictive value:

$$\text{PPV} = \text{true positives/total number of positive test results} = a/(a+b)$$

The diagnostic specificity is defined as

$$\text{specificity} = \text{true negatives/total nondiseases} = d/(b+d)$$

Specificity is the ability to correctly identify those without the disease (true negative rate). It is different from the negative predictive value:

$$\text{NPV} = \text{true negatives/total number of negative test results} = d/(c+d)$$

Another term, *false positive rate* is often used:

$$\text{false positive rate} = (1\text{-specificity}) = b/(b+d).$$

The optimal sensitivity and specificity are 100%: these are perfect tests without false-positive or false-negative results. In diagnostic endoscopy, the relative sensitivity and specificity changes as compared to a conventional diagnostic method on the same study population are more meaningful performance measures for a new diagnostic method. The reason is that in any clinical study, we cannot guarantee that we examined all the diseases. Theoretically, for one to get all the diseases, the whole organ of the patient has to be sectioned for histopathology examination, which is obviously impractical.

To fully characterize a diagnostic method, the sensitivity and specificity variations have to be examined by varying the decision threshold. The resulting plot is known as the ROC curve (Swets 1988). Figure 1.1 shows a couple of sample ROC curves. When the strictest decision threshold is applied, all cases are classified as diseases, and the ROC curve is at the upper right corner of the plot (100% sensitivity, 0% specificity). As the threshold is loosened, sensitivity decreases, and specificity

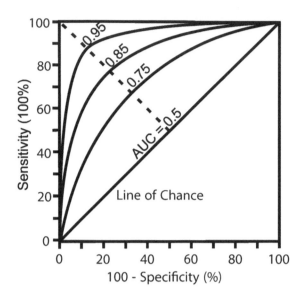

FIGURE 1.1   The sample ROC curves demonstrate the trade-off between sensitivity and specificity as the test is operated at different points along the curve. (Reproduced with permission from Swets, J.A. 1988. *Science* 240: 1285–1293.)

increases until all cases are classified as nondiseases, and the ROC curve reaches the lower left corner of the plot (0% sensitivity, 100% specificity). The closer the ROC curve to the upper left corner (gold standard: 100% sensitivity, 100% specificity), the better the test. The area under the ROC curve (AUC) can be used to compare tests. The sample ROC curves in Figure 1.1 vary in AUC from 0.95, 0.85, 0.75 to 0.5 (the line of chance). For a given test, one can choose to operate at different points on the ROC curve for different clinical purposes.

1.3   LIGHT-TISSUE INTERACTION: INITIAL EVENTS

When a beam of light reaches the tissue surface (Figure 1.2, bronchial tissue as an example), part of it will be reflected by the surface directly, while the rest will be refracted and transmitted into the bronchial tissue. The direct reflection by the bronchial surface is called *specular reflection*, and is related only to the refractive index change between air and the epithelium of the bronchial tissue. The light transmitted into the tissue will be scattered and absorbed by the tissue. After multiple scattering, some of the transmitted light will reemerge through the bronchial surface into the air. This reemergence is called *diffuse reflection*. The amount of diffuse reflection is determined by both scattering and absorption properties of the bronchial tissue. The stronger the absorption, the less the diffuse reflection; the stronger the scattering, the large the diffuse reflection. Within the reflected photons, there are also a very small portion that has only been scattered once in the tissue, the so-called *single scattering photons*. Following absorption of a photon by the bronchial tissue, an electrically excited absorbing molecule may rapidly return to a more stable energy state by

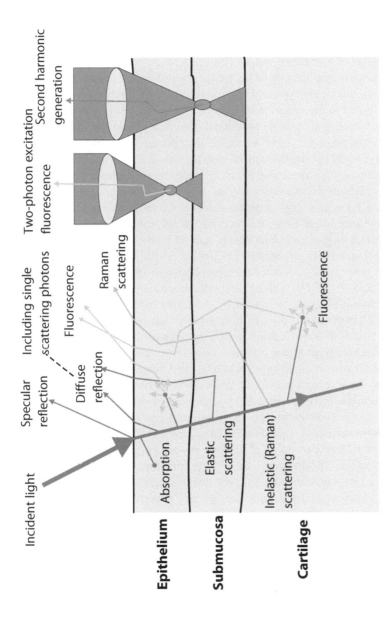

FIGURE 1.2 (See color insert) Schematic diagram of light pathways in bronchial tissue. A beam of incident light could interact with the bronchial tissue and generate various secondary photons measurable at the tissue surface: specular reflection, diffuse reflection, fluorescence, Raman scattering, two-photon excitation fluorescence, and second harmonic generation. These measurable optical properties can be used for determining the structural features as well as the biochemical composition and functional changes in normal and abnormal bronchial tissues. (Reproduced with permission from Zeng, H. et al. 2004. *Photodiagnosis and Photodynamic Therapy* 1: 111–122.)

reemission of a photon with lower energy, that is, *fluorescence emission*. Fluorescence photons are also subjected to tissue reabsorption and scattering before some of these photons reach the tissue surface. Most light scattering in tissue is *elastic scattering* (or Rayleigh scattering) with no change in photon energy (or frequency). A very small portion of the scattered light, about 1 in 10^{10}, is *inelastically scattered* (Raman scattering) with a corresponding change in frequency. The difference between the incident and scattered frequencies corresponds to an excitation of the molecular system, most often excitation of vibrational modes (Figure 1.3). By measuring the intensity of the scattered photons as a function of the frequency difference, a Raman spectrum is obtained. Raman peaks are typically narrow (a few wavenumbers) and in many cases can be attributed to the vibration of specific chemical bonds (or normal mode dominated by the vibration of a single functional group) in a molecule. As such, it is a "fingerprint" for the presence of various molecular species. If ultrafast pulsed light (femtosecond pulse width) is tightly focused into the tissue, two or multiple photons can work together to excite molecules in tissue, generating nonlinear effects, such as *two-photon excitation fluorescence, second harmonic generation* (right side of Figure 1.2). Figure 1.3 is an energy diagram showing the molecular processes of all the previously mentioned light-tissue interactions, namely, absorption, fluorescence, two-photon excitation fluorescence (nonlinear), second harmonic generation (nonlinear), elastic scattering, and Raman scattering (Stokes and Anti-Stokes).

Example light absorption molecules (chromophores) in internal epithelial tissue includes proteins and DNA (UV), hemoglobin (visible wavelength), and water (infrared). Elastic scattering in tissue is caused by refractive index fluctuations seen at cell membrane, cell nuclei, and collagen bundles, etc. Various proteins, lipids, and nucleic acids in tissue all show Raman scattering. Fluorescent molecules (fluorophores) in tissue include tryptophan, tyrosine, porphyrins, collagen, elastin, flavins, and nicotinamide adenine dinucleotide (NADH), etc. Collagen is an example molecule that can produce second harmonic generations in tissue. Thus, the various light-tissue interactions and tissue optical properties can be used for determining the structural features as well as the biochemical compositions and functional changes in normal and abnormal tissues. Based on the initial events of light-tissue interaction outlined in Figure 1.2, we can perform various endoscopy imaging and spectroscopy, such as reflectance spectroscopy, white light imaging, fluorescence spectroscopy, fluorescence imaging, confocal imaging (using single scattering photons and fluorescence from exogenous dyes), optical coherence tomography (OCT, using single scattering photons as well), and multiphoton excitation imaging. This book will show how these imaging and spectroscopy modalities are implemented and how they help improve clinical diagnoses.

1.4 OPTICAL ENDOSCOPY IS COMPLEMENTARY TO TRADITIONAL MEDICAL IMAGING FOR CLINICAL DIAGNOSIS

X-ray imaging (including CT) and magnetic resonance imaging (MRI) are the two major traditional physical diagnostic methods in clinical use. Other methods include

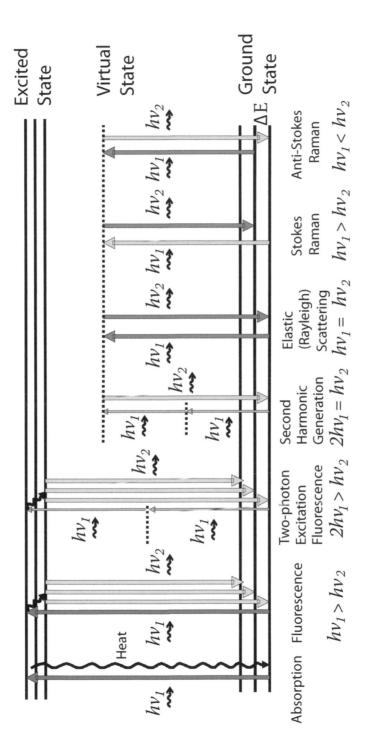

FIGURE 1.3 Energy diagram showing the molecular processes associated with various light-tissue interactions illustrated in Figure 1.2.

ultrasound and positron emission tomography (PET), etc. (Profio 1993). Examination of the underlying physical principles shows that x-ray imaging only gives the element distribution in tissues. The high energy x-ray absorption by tissue is caused by striking out the inner shell electrons of an atom and therefore, is not sensitive to the chemical state (outer shell electronic states) of the molecules. The MRI exploits the nuclear spin to form an image of certain nuclides (i.e., ^{1}H) distribution in tissue. MRI is capable of deriving chemical information and has many advantages over x-rays. Optical diagnostic methods are currently being intensively studied. The light absorption and fluorescence emission are determined by the outer shell electronic states of an atom in its compound. Vibrational methods (Raman scattering) also deals with chemical bonds between atoms. Therefore, using optical methods, we detect or image the chemical states of the biological tissue. Optical spectroscopy and imaging are becoming complimentary methods of diagnoses to x-ray and MRI. They may be more powerful in detection of premalignant lesion or detection of a lesion in its early stage before the element redistribution becomes visible under x-rays or MRI. Optical endoscopy methods enable visualization of internal organ surfaces by a physician and also enable optical diagnosis of internal organs. Optical methods are more suitable for surface lesion detection, while traditional medical imaging modalities are excellent for imaging body anatomy. There are unique opportunities for optical methods and endoscopy for early cancer detection because more than 80% of cancers arise from the surface of epithelial tissues such as the skin, lung, bronchial tree, gastrointestinal tract, cervix, and ENT (ear, nose, and throat). And even most breast cancers are started from the inner surface of milk ducts, which can be accessed by very thin endoscopes.

1.5   COMPLIMENTARY ROLES OF VARIOUS ENDOSCOPY MODALITIES: ORGANIZATION OF THIS BOOK

Figure 1.4 outlines the endoscopy imaging and spectroscopy modalities to be covered by this book. It is organized by the modes of illumination and detection. Broad-beam illumination (beam diameter larger than 10 mm) is usually used for *field imaging* modalities that can survey the whole internal organ surface rapidly to identify abnormal regions or certain structures for further detailed measurements (point spectroscopy and "point" microscopy), or for biopsy and other interventions. Of the listed field imaging modalities, white light endoscopy is essential for visualizing the internal organ surface and for guiding other measurements and interventions. Fluorescence imaging has the highest diagnostic sensitivity, but poor specificity for cancer detection. *Point spectroscopy* (reflectance and fluorescence) can also be performed under broad beam illumination through controlled point detection.

Narrow beam illumination (beam diameter smaller than 1 mm) is often used for catheter-based *point spectroscopy* modalities. Reflectance, fluorescence, and Raman spectroscopies can all be carried out through fiber optic catheters. Spectroscopy is capable of providing high diagnostic specificity. However, the diagnostic sensitivity is largely limited by the field imaging modality used to guide the spectral measurement since it is impossible to perform point spectroscopy over the whole field.

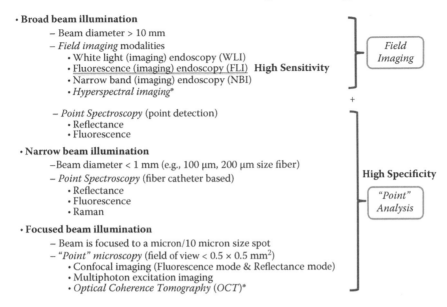

Different Modalites in Diagnostic Endoscopy

FIGURE 1.4 Different modalities in diagnostic endoscopy organized by modes of illumination and detection.

Focused beam illumination is employed for microscopy modalities. The light beam is focused to a 10 micron spot inside the tissue for OCT and less than 1 micron for confocal and multiphoton microscopy. However, the field of view is usually less than 0.5×0.5 mm^2 (OCT could be larger). So it is usually still a "point" measurement (<1 mm size). We will call it *"point" microscopy* throughout the book. Endoscopy microscopy provides histology-like images, thus, could have high diagnostic specificity. Similar to point spectroscopy, the field imaging guidance determines the diagnostic sensitivity.

In Figure 1.4, a star mark is put besides hyperspectral imaging and OCT because these two modalities have the potential to serve as both fielding imaging and point spectroscopy or "point" microscopy. With dramatic improvement on data acquisition speed, hyperspectral imaging can one day provides image of a large field at video rate and simultaneously a reflectance spectrum from each pixel. An advanced OCT system can obtain sequential sectional images of a lumen very rapidly and reconstruct 3-D images of the lumen as it scans along. Further imaging speed and imaging quality improvements are still required for practical clinical applications for OCT as a field imaging modality.

It is hard for this book to cover every optical imaging and spectroscopy technique that can potentially be used for endoscopy applications. Some of these imaging modalities, currently at the initial development stage for endoscopy, such as CARS (coherent anti-Stokes Raman scattering), SRS (stimulated Raman scattering), and photoacoustic imaging, could one day become important modalities for diagnostic endoscopy.

The 13 chapters in this book are divided into five sections. Section I is about the basics of endoscopy and light-tissue interactions (Chapters 1–3). Section II covers endoscopic field imaging modalities (Chapter 4—white light endoscopy and Chapter 5—fluorescence endoscopy). Section III deals with endoscopic point spectroscopy (Chapters 6 and 7), covering reflectance, fluorescence, and Raman. Section IV is dedicated to endoscopic "point" microscopy (Chapters 9–11), including confocal microscopy, OCT, multiphoton excitation imaging, and micromanufacturing technology for catheter development. Section V presents clinical applications in the lung (Chapter 12) and the gastrointestinal tract (Chapter 13).

Through this book, we hope to convince the readers that optimized future diagnostic endoscopy systems should include at least a high diagnostic sensitivity, field imaging modality and a high diagnostic specificity, point analysis modality, as their essential components.

REFERENCES

Benedict, E.B. 1951. *Endoscopy*. Baltimore: The Williams & Wilkins Company.

Berci, G. (ed.), 1976. *Endoscopy*. New York: Appleton-Century-Crofts.

Cotton, P.B. and Williams, C.B. (eds.) 2003. *Practical Gastrointestinal Endoscopy—the Fundamentals*. Oxford: Blackwell Publishing.

Hirschowitz, B.I., Curtis. L.E., Peters, C.W., and Pollard, H.M. 1958. Demonstration of a new gastroscope: "The fiber-scope." *Gastroenterology* 35: 50–53.

Hopkins, H.H. and Kapany, N.S. 1954. A flexible fiber scope using static scanning. *Nature* 173: 39–41.

Nelson, R.S. 1970. *Endoscopy in Gastric Cancer*. Berlin, Heidelberg, New York: Springer-Verlag.

Profio, A.E. (ed.) 1993. *Biomedical Engineering*. New York: John Wiley & Sons.

Swets, J.A. 1988. Measuring the accuracy of diagnostic systems. *Science* 240: 1285–1293.

Van Heel, A.C.S. 1954. A new method of transporting optical images without aberration. *Nature* 173: 39.

Zeng, H., McWilliams, A., and Lam, S. 2004. Optical spectroscopy and imaging for early lung cancer detection: A review. *Photodiagnosis and Photodynamic Therapy* 1: 111–122.

2 Basic Components of Endoscopes

Haishan Zeng
British Columbia Cancer Agency Research Centre,
Vancouver, British Columbia, Canada

CONTENTS

An endoscope often consists of a rigid or flexible tube, a light delivery system, an image relay system, and in many cases, an instrument/biopsy channel. To support the proper functioning of an endoscope, dedicated electronics and mechanical parts, water, and air supply are also part of a complete endoscopy system. In a rigid tube endoscope, rod lenses, prisms, and other rigid optical components are used to relay both the illumination light and the image. It is often called a *rigid endoscope*. Rigid endoscopes are often used in shallow and easy-to-access body cavities such as the nose and throat. It is also used for normally closed body cavities through a small incision such as the abdominal or pelvic cavity (laparoscope) and the interior of a joint (arthroscope).

A flexible tube endoscope is called a *flexible endoscope*. Usually a flexible optical fiber bundle is used to transmit the illumination light, while another fiber bundle is used to relay the image formed by the objective lens at the distal end to the proximal end eyepiece for direct visualization or for interface with a video camera. This type of flexible endoscope is called a *fiberoptic endoscope*. Images can also be captured by placing a miniaturized CCD chip (charge coupled device) at the distal end of the endoscope. In this case, the images are relayed electrically and displayed directly on a viewing monitor. This type of endoscope is called a *video endoscope*. Flexible endoscopes are often used in hard-to-access body cavities and lumens such as the gastrointestinal tract, the respiratory tract, and the urinary tract.

2.1   MECHANICAL STRUCTURES OF AN ENDOSCOPE

Figure 2.1 show the mechanical structures of an endoscope using a fiberoptic colonoscope as an example. A colonoscope is used to visualize the colon and rectum and has the most complicated structure among all endoscopes. The distal tip contains the illumination optics, the imaging objective, an instrument/biopsy channel, and the mechanics for providing air, water, and suction. The bending section contains pulling wires, passing back through the length of the instrument shaft to the angulation control at the proximal head and enabling the tip to deflect at large angles as high as 180° degrees and in four directions. The proximal head of the endoscope contains the eyepiece optics, the entry into the instrument/biopsy channel, and angulation control knobs, air/water and suction valves. An umbilical cord connects the proximal head to the light source to obtain illumination light through a light guide and to transmit air, water, and suction through other tubes. The length of a colonoscope could be as long as 2 meters.

The distal tip of an endoscope is shown in Figure 2.2, illustrating the locations of the imaging objective lens, the illumination channel, the air/water channel, and the instrument channel. Two illumination channels are located on either side of the objective lens to more uniformly illuminate the imaging area. A large instrument channel (usually 2–4 mm in diameter) allows the passage of fine flexible accessories (e.g., biopsy forceps, cytology brushes, therapeutic devices) from a port on the endoscope proximal head (Figure 2.1) through the shaft and into the field of view. It also facilitates suction and the removal of tissue (thus, sometimes called a *biopsy channel*). The many spectroscopy analysis catheters and microscopy imaging catheters to

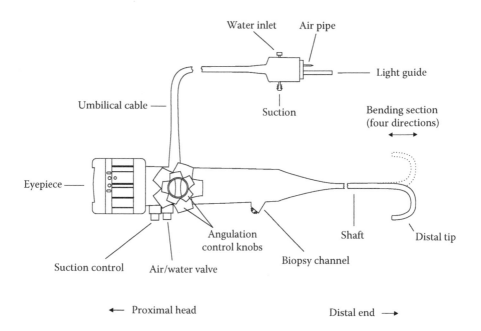

FIGURE 2.1   Schematics showing basic mechanical structures of an endoscope.

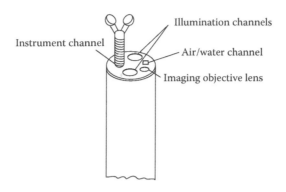

FIGURE 2.2 The distal tip of an endoscope.

be discussed later in this book are also delivered through this channel. Therapeutic endoscopes sometimes have twin channels to facilitate more efficient operations. The air/water can transmit air to insufflate and expand the organ being examined. The air is supplied from a pump usually located inside the light source and is controlled by the air/water valve (Figure 2.1) in the endoscope proximal head. The air system also applies pressure to a water bottle so that a jet of water can be squirted across the lenses to clean them. In colonoscopies there is an additional proximal opening for the air/water channel to facilitate high-pressure flushing with a syringe.

The insertion tube contents are schematically shown in Figure 2.3. It contains the illumination fiber bundle, the imaging fiber bundle, the angulation wires, the air pipe, the water pipe, and the instrument channel. For mechanical protection, these functional components are wrapped by intertwined metal bands and then covered with metal wire mesh along the length of the tube. The outside covering tube is a durable plastic sheath capable of withstanding caustic body fluids and also disinfectants used in cleaning the endoscopes. For a new type of *variable stiffness* colonoscope (Innoflex®, Olympus Optical Co.), an extra steel coil is built into the insertion tube that can be expanded or contracted as required through a control knob to vary the degree of flexibility and stiffness. This permits the insertion tube

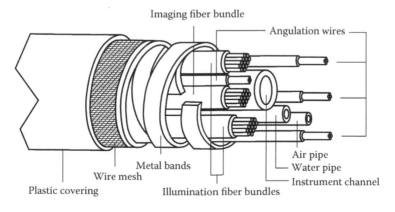

FIGURE 2.3 Schematic representation of the insertion tube of an endoscope.

to be adjusted according to an individual patient's own anatomical conditions and contours of the colon.

2.2 ELECTRONICS IN AN ENDOSCOPY SYSTEM

Figure 2.4 shows an example of a complete flexible video endoscope system. It consists of the light source, the video endoscope, the video processor, and the video monitor. Electronics involved include controlling electronics for the CCD detector located right behind the objective at the distal tip, electronics in the light source to provide power supply to the lamp (usually a high pressure Xenon arc bulb) and to control the air/water supply, imaging acquisition, and processing electronics in the video processor, and display electronics in the monitor. For a fiberoptic endoscope based system, a color CCD video camera is often connected to the endoscope eyepiece for image acquisition and electronic display on a monitor, replacing the traditional visual observation through the eyepiece using the endoscopist's naked eye.

Besides providing electrical power supply to the lamp, the electronics in the light source also perform other duties and functions, such as light-level control, storing illumination settings, air/water pump control, and driving cooling fans. Electronics in a light source for a video endoscope may also involve controlling a rotating filter wheel to provide sequential red (R), green (G), blue (B) illumination and provide synchronization signals to the CCD control electronics.

Electronic white light color images for endoscopy can be obtained in two different ways: the sequential RGB imaging method and Bayer filtering method. In the *sequential RGB imaging* method as illustrated in Figure 2.5, a rotating filter wheel is used in the light source to generate blue (B: 400–500 nm), green (G: 500–600 nm), and red (R: 600–700 nm) colored light sequentially, while a CCD detector sensitive to all visible wavelengths (at least covering 400–700 nm) is used to capture the R, G, and B images sequentially. To achieve video rate imaging, the wheel is rotated at 25 or 30 Hz according to different video standards. There are opaque blocks between the three filters to prevent mixed color light exposure and sequential RGB imaging data readout can be performed during these time periods. Three memory boards

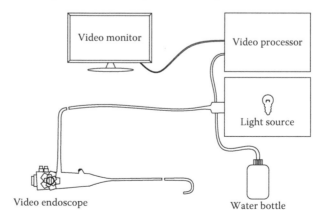

FIGURE 2.4 An example endoscopy system.

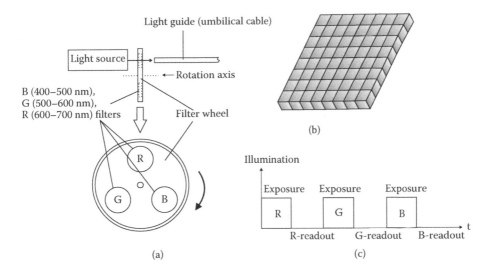

FIGURE 2.5 Principle of sequential RGB imaging. (a) sequential R, G, B illumination; (b) CCD detector; (c) Timing series of R, G, B exposure and image readout.

may be used to store the three R, G, and B images. The digital images could then be converted into video signals for image display on a video monitor. This imaging method is often used in video endoscopy, facilitating a smaller CCD detector that helps miniaturization or gives higher resolution.

In the *Bayer filtering* method as illustrated in Figure 2.6, a single hot mirror filter (instead of the filter wheel in Figure 2.5) is used to simply generate a broad band white light (400–700 nm) illumination, while color imaging is realized by placing a Bayer filter mosaic on top of the CCD array. In a Bayer filter mosaic, each quartet of pixels utilizes three different filters for color separation. To match the human eye response one pixel filter transmits only red light, one pixel transmits only blue light, and two transmit only green (see Figure 2.6b). This filter overlays the CCD pixels with one-to-one correlation; in other words, when the detector captures an image,

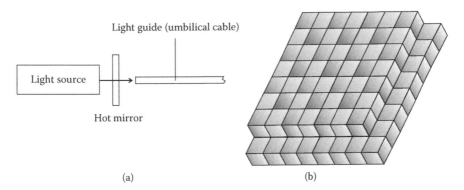

FIGURE 2.6 (See color insert) Principle of Bayer filtering color imaging. (a) simplified white light illumination; (b) Bayer filter mosaic on top of a CCD array.

simultaneously 25% of the pixels capture red wavelengths, 25% capture blue wavelengths, and 50% capture green wavelengths. The resultant data is processed using color-space interpolation algorithms to create an RGB color image. In this imaging method, the resolution is degraded or a larger CCD detector is required to keep the same resolution. The advantages are simplicity and reduced cost for the complete endoscopy system. It is often used for electronic image acquisition and display in a fiberoptic endoscopy system, but could be used in video endoscopy as well.

2.3 BASIC OPTICS IN ENDOSCOPY

The basic optics involved in endoscopy includes illumination optics and imaging optics. Optical components used are lens, optical fibers, and prisms. Lenses perform imaging functions or collecting light for illumination. Optical fibers can be used for both illumination light transmission and relaying images. Prisms are for changing the image viewing directions or illumination light direction.

2.3.1 FIBER OPTICS

An optical fiber guides light on the basis of a step index structure that causes total internal reflections and directs the coupled light almost without loss from one end to the other. For this to happen the refractive index of the fiber core material (n_1) has to be larger than that of the fiber cladding material (n_2). Figure 2.7 illustrates how incident light rays within the acceptance cone can be guided through the fiber by total internal reflection, while light with an incidence angle larger than the maximum acceptance angle θ_a will be partially refracted outside of the fiber. The remaining partially reflected light will eventually get lost after multiple interactions with the fiber core-cladding interface. The maximum acceptance angle θ_a is determined by the condition for total internal reflection at the core-cladding interface:

$$\sin(\theta_c) = \frac{n_2}{n_1} \tag{2.1}$$

Assuming the fiber is in the air with a refractive index of 1, the following holds for the entry surface:

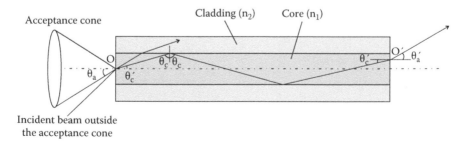

FIGURE 2.7 Light paths inside an optical fiber.

$$\sin(\theta_a) = \frac{n_1}{1} \cdot \sin(\theta_c') \tag{2.2}$$

With $\theta_c' = 90° - \theta_c$, this yields the numerical aperture (N.A.) for the fiber:

$$\text{N.A.} = 1 \cdot \sin(\theta_a) = n_1 \cdot \cos(\theta_c) = n_1 \cdot \sqrt{1 - \sin^2(\theta_c)} = n_1 \cdot \sqrt{1 - n_2^2/n_1^2} = \sqrt{n_1^2 - n_2^2} \tag{2.3}$$

Another important characteristics of light transmission through an optical fiber is that light rays entering and exiting the fiber keep the same angle with respect to the fiber's central axis. Figure 2.7 illustrated this situation for a straight fiber. The incident light ray shown at the left side of the drawing makes an angle, θ_a, with the fiber's central axis. It is refracted at point O and enters the fiber at an angle, θ_c', to the fiber's axis determined by:

$$\sin(\theta_c') = \sin(\theta_a)/n_1 \tag{2.4}$$

The light ray inside the fiber encounters multiple times of total internal reflections and exits at point O′ on the exit face of the fiber. The light ray incident angle at point O′ is the same as θ_c'. The light ray exit angle, θ_a', is determined by the law of refraction as follows:

$$1 \cdot \sin(\theta_a') = n_1 \cdot \sin(\theta_c') \tag{2.5}$$

Comparing Equation (2.4) and Equation (2.5), we conclude that $\theta_a' = \theta_a$, i.e., the entrance angle and the exit angle are kept the same. Hopkins has demonstrated that this is also true for a curved fiber (Hopkins 1976).

Optical fibers used in an endoscope are required to transmit a broad wavelength range of light (at least covering the visible wavelength range 400–700 nm), therefore, have to be the "multi-mode" type and have a core diameter of larger than about 8 µm. More than ten thousands of individual glass fibers form fiber bundles of a few mm in size and are used to transmit illumination light or relay the image in an endoscope. For illumination, fibers with a diameter of larger than 25 µm are used and there is no need to have any precise order in the bundle (*incoherent bundle*). For relaying an optical image in an endoscope, fibers of smaller diameter (~10 µm size) are used for better image resolution and the fiber bundle must be ordered. Each fiber has to occupy the same relative position at the exit face of the bundle as at the entry face. This is known as a *coherent bundle*. These imaging bundles are often used to make flexible endoscopes.

There is a special type of single optical fiber that can also be used to relay optical images. In this special fiber, the refractive index of the core material varies. It has the highest refractive index at the center and decreases gradually towards the edge following a parabolic profile across the radius of the fiber as illustrated in Figure 2.8. This is called the *gradient index (GRIN) fiber* or *Selfoc® (self-focusing) fiber*. This

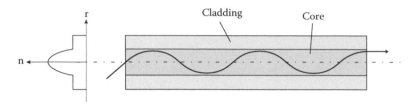

FIGURE 2.8 Light transmission in a Selfoc® (self-focusing) fiber, which has a gradient refractive index in the core (also called GRIN fiber). A single Selfoc® fiber is used for image relay in some rigid endoscopes.

fiber can periodically focus an image along its length. Therefore, depending on where the end face is cut on the fiber, a Selfoc® fiber can act as a converging lens, a plane-parallel plate (with or without inversion), or a diverging lens. To form an endoscope, two different types of Selfoc® fibers are utilized: one act as the objective lens, the other as the relay lens. The objective lens Selfoc® fiber is much shorter than the relay lens Selfoc® fiber and is bonded to the end of the relay lens Selfoc® fiber. The objective Selfoc® fiber images an object outside of the distal end of the endoscope onto its rear face, while the relay lens Selfoc® fiber pass on this image to the proximal end of the endoscope. These endoscopes have smaller diameters and facilitate higher resolution imaging, but are generally rigid because the Selfoc® fibers are very fragile.

2.3.2 ILLUMINATION OPTICS

The illumination optics consists of three parts: light collection from a lamp, light transmission through a flexible light guide, and illumination lenses at the distal end. Figure 2.9a shows the complete illumination path for a rigid endoscope, Figure 2.9b,c are example variations of the collection optics. In Figure 2.9a, light collection from the lamp is performed by a large collection angle lens (the collection lens). A spherical reflection mirror placed behind the lamp help in collecting light rays going backwards, i.e., light from both the solid angle α and α' are collected by the collection lens. This roughly doubles the total power getting into the light guide. A focus lens then focuses the collected light to the flexible light guide. A heat absorbing or heat reflecting filter (heat filter) is usually placed between the lamp and the collection lens to protect the more dedicated optical components afterward and the light guide. Another optical filter can be placed between the collection lens and the focus lens, where the light beam is roughly collimated, to further refine the spectral profile of the light source to suit the intended imaging modality. For example, for white light endoscopy, the illumination spectrum is from 400–700 nm.

There are two common variations to the light collection optics. Figure 2.9b shows that a parabolic mirror is used to collect and collimate light from the lamp, which is located at the focal point of the parabolic. A focus lens then focuses the collimated beam into the light guide. The heat filter and the spectral shaping filter are placed directly between the lamp and the focus lens. The parabolic mirror is often a dichroic type that reflecting the visible wavelengths and transmitting the infrared (IR) wavelengths. It is called a *cold mirror* and also helps reduce the heat to the light

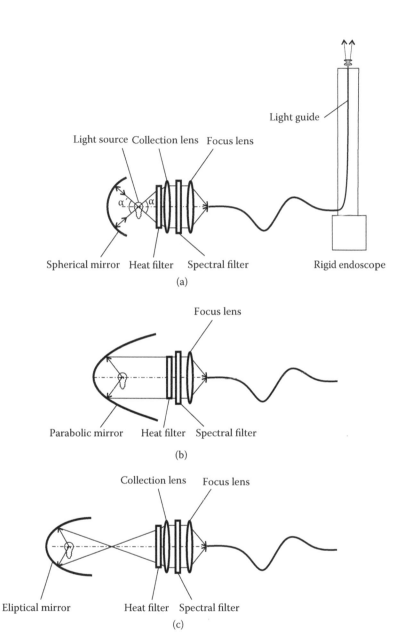

FIGURE 2.9 Illumination optics for endoscopy. (a) The complete illumination path for a rigid endoscope, (b) one variation of the light collection optics using a parabolic mirror, (c) another variation of the light collection optics using an elliptical mirror.

guide. Figure 2.9c shows that an elliptical mirror is used in light collection. The lamp is located at one of the two foci of the elliptical mirror, while the collected light is focused to the other focal point of the elliptical mirror. A collection lens is placed after the focused beam to accept and collimate the light beam and a focus lens then focuses the collimated beam to the light guide. Similar to Figure 2.9a, the heat filter is placed before the collection lens, while the spectral shaping filter is placed between the collection lens and the focus lens. The elliptical mirror is also often a dichroic type. Usually the parabolic mirror and elliptical mirror help collect light from a larger solid angle than the spherical mirror configuration.

In the early days, tungsten lamps were used for endoscopy illumination. Now high-pressure Xenon arc lamps and metal halide lamps are used, which have a more flat spectral profile and are brighter with a smaller arc size, making it easier to focus to the light guide.

For a flexible endoscope (both the fiber endoscope and video endoscope), a single light guide based on an incoherent fiber bundle is used to conduct light from the light source to the distal end of the endoscope. This light guide has a diameter of millimeters with an individual fiber size of 25 µm or greater. From the light source to the endoscope proximal end, the light guide is packed within an umbilical cord that also contains pipes to transmit air and water.

For a rigid endoscope with no requirements for air and water transmission, the light guide takes the form of a two segments design. Usually a flexible liquid light guide is used between the light source and the proximal end of the endoscope. A mechanical adapter, allowing the liquid light guide to be detachable, then helps connect the liquid light guide with an incoherent fiber bundle-based light guide inside the endoscope to transmit the light to the distal end. A liquid light is used in the first segment for its higher light transmittance than a fiber bundle.

At the distal end of the fiber bundle, a set of lenses is used to achieve uniform and large field illumination. In endoscopy applications, the object being viewed is usually not flat, but a curved surface. Therefore, the use of regular aberration-free projection lenses does not lead to uniform illumination; instead, aspheric lenses are employed to achieve uniform illumination in endoscopy. Miniaturization requirements deem these lenses to have about the same millimeter scale sizes of the illumination fiber bundle.

2.3.3 IMAGING OPTICS

The imaging optics for an endoscope could include the objective, the image relay optics, and the eyepiece. Optical design for a rigid endoscope is the most challenging, where image relay is accomplished by as many as 60 individual lenses. A fiber optic flexible endoscope uses a coherent fiber bundle for image relay, simplifying the optics considerably. In the case of video endoscope, images are captured by placing a miniaturized CCD sensor directly after the objective, eliminating the uses of image relay optics and the eyepiece.

Figure 2.10 illustrates the imaging optics involved in a fiber optic endoscope. Object QP on the tissue surface is imaged by the objective to form an image Q′P′ at the entry-face of the distal end of the coherent fiber bundle. The coherent bundle,

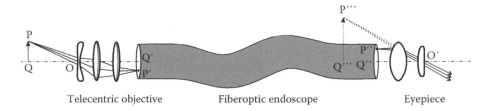

Telecentric objective   Fiberoptic endoscope   Eyepiece

FIGURE 2.10 Image formation and relay in a fiber optic endoscope.

containing several hundred thousand fibers, relays the image to the proximal end-face of the bundle to form an image Q″P″. The eyepiece then create a virtual image Q‴P‴ that presents a magnified view of the object to the observer's eye. Optimally the objective should be designed to be telecentric in the image space as shown in Figure 2.10. With the telecentric objective the chief (central) rays of the light cones forming the image Q′P′ all enter normally on the entry-face of the bundle. This is illustrated for the cone of rays forming the image at P′ of the object point P. Because light rays entering and exiting an optical fiber in the bundle keeps the same angle with respect to the fiber central axis (Figure 2.7), the chief rays of the light cones emerging from the proximal end-face of the fiber bundle are then also perpendicular to the end-face of the bundle. This is exemplified for the cone of light emerging from the fiber at P″ (the relayed image of the object point P). The chef rays of all emerging cones of light from image Q″P″ are thus parallel to the axis, and they all therefore pass through the back focal point (O′) of the eyepiece. This guarantees that all other rays from each cone can pass the exit-pupil of the system, therefore, achieves uniform detection efficiency over the whole image. If an ordinary objective is used, the rays for any off-axis object point (such as P) emerge from the eyepiece as an annular tube of rays. For details, see (Hopkins 1976). Some rays will fall outside of the exit-pupil, resulting in nonuniform image detection efficiencies (smaller on the edge than in the center).

Figure 2.11 shows the imaging optics of a rigid endoscope consisting of the objective, the field lenses, the relay lenses, and the cyepiece. The objective is designed to be telecentric in the image space to facilitate interface with the relay lenses. The objective creates an inverted image Q′P′ of the internal organ QP. Field lens 1 located close to image Q′P′ then redirects the light cones towards the relay lens 1 with the chief rays pass through the center of relay lens. Field lens 2 placed near image Q″P″ helps maintain the chief rays confined within the small tube of the endoscope. Relay lens 2 then conveys the image to Q‴P‴. This series of field lenses and relay lenses are repeated inside the endoscope tube until the image is relayed to the eyepiece, which presents a final magnified image for viewing by the physician. Figure 2.11b shows that in a modern rigid endoscope the conventional achromatic doublets-based relay lens is replaced by Hopkin's rod-lens. One advantage of using rod-lens is that the light beam can be confined more tightly to the optical axis by reducing the ray divergence in air gaps between the lenses. This reduces vignetting. Another advantage is that rod-lenses are simpler to mount and have larger clear apertures than thin achromatic lenses.

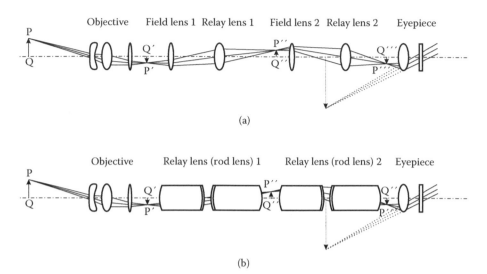

FIGURE 2.11   Imaging optics of a rigid endoscope. (a) Optical layout of a conventional rigid endoscope, (b) modern rigid endoscope based on Hopkin's rod-lenses for image relay.

Nowadays, for both the rigid endoscope and fiber optic endoscope, a CCD video camera is often attached to the eyepiece to capture and display the video images on a monitor in real-time. For this purpose a lens adapter is used between the eyepiece and the CCD camera to project the image to the CCD sensor.

2.4   NEW DEVELOPMENT—WIRELESS CAPSULE ENDOSCOPY

Recently a new type of endoscopy have been developed, which has completely different configurations from traditional endoscopy (Iddan 2000; Moglia 2008). It involves a wireless capsule endoscope and facilitates for the first time painless endoscopic imaging of the whole small bowel. A wireless capsule endoscope utilizes miniaturized opto-electronic mechanical components for illumination, imaging, power supply, and telemetry. Figure 2.12 shows the schematics of such a wireless capsule endoscope. It consists of a transparent optical dome (1), an objective lens (2), four or six white light LEDs (light emitting diode) (3), a CMOS (complementary Metal Oxide Semiconductor) or CCD image sensor (4), silver oxide batteries (5), and an ASIC (application specific integrated circuit) (6). The ASIC is a key component that controls the LED for illumination, the CMOS or CCD sensor for image acquisition, and also includes a UHF-band radio-telemetry to transmit the video images to an array of antennas placed externally at the patient's abdomen and stored into a portable recorder. All these components are mounted inside a capsule of about 11 mm diameter and 26–31 mm length, small enough to transport passively through the digestive tract by peristalsis.

Current applications of capsule endoscopes are solely for the digestive tract, especially for the small intestine, which cannot be fully accessed by traditional endoscopes. For esophagus examination, it takes 14 images/sec for a total period of 15–20

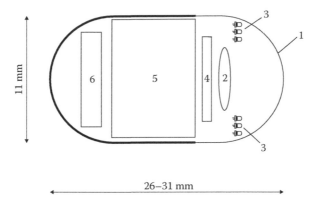

11 mm

26–31 mm

FIGURE 2.12 Schematics of a wireless capsule endoscope.

minutes. For the colon, it takes images at 4 frames/sec for a total period of about 3 hours. For the small intestine, images are acquired at 2 frames/sec for 8 hours due to the slow and long journey (20–25 feet) of the capsule. The device was found to be clinically useful for the detection of GI bleeding, Crohn's disease, celiac disease, and small bowel tumors. It also showed encouraging results for detecting Barrett's esophagus and screening colorectal diseases. However, current capsule endoscopes do not have any intervening functions of traditional endoscopes, such as performing biopsies and facilitating therapeutic procedures.

REFERENCES

Hopkins, H.H. 1976. Chapter 1: Optical principles of the endoscope and Chapter 2: Physics of the fiberoptic endoscope. In *Endoscopy*, ed. G. Berci. New York: Appleton-Century-Crofts, pp. 3–63.

Iddan, G., Meron, G., Glukhovsky, A., and Swain, P. 2000. Wireless capsule endoscopy. *Nature* 405: 417.

Moglia, A., Menciassi, A., and Dario, P. 2008. Recent patents on wireless capsule endoscopy. *Recent Patents on Biomedical Engineering* 1: 24–33.

3 Light Propagation in Tissue—Radiative Transport Theory

Haishan Zeng
British Columbia Cancer Agency Research Centre,
Vancouver, British Columbia, Canada

CONTENTS

Biological tissues are highly scattering, turbid media. This makes it difficult to predict the light pathways in tissue and renders optical imaging inside the tissue a challenging task. Figure 3.1 schematically shows the different pathways when a beam of light shines on a transparent medium, such as a piece of glass, versus a turbid medium, such as a tissue slab. For the glass piece, its optical properties can be completely described by the refractive index n and the absorption coefficient μ_a. Light transport and pathways can be easily predicted by geometry optics. For the tissue slab, due to scattering two more optical parameters, the scattering coefficient μ_s and the scattering anisotropy g are added to its optical properties description. Light transport and pathways become difficult to predict due to multiple scattering. Also shown are the fluorescence escape process that is complicated by multiple scattering as well.

Historically, the problem of wave propagation in turbid media has been investigated theoretically from two distinct points of view (Ishimaru 1978). One is "radiative transfer theory" or "transport theory," and the other is "multiple scattering theory" or "analytical theory." Analytical theory starts with basic differential equations such as the Maxwell equation, obtains solutions for a single particle, introduces the interaction effects of many particles, and then considers statistical averages. It is mathematically rigorous in the sense that in principle, all the multiple scattering, diffraction, and interference effects can be included. However, practically, it is almost impossible to obtain the exact properties of all tissue components. It is mathematically difficult to treat their multiple scattering interactions because tissue

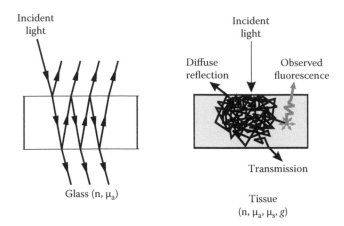

FIGURE 3.1 Schematic diagram comparing light transport in a transparent medium (glass) versus in a turbid medium (tissue).

is irregularly shaped, inhomogeneous, multi-layered, and has anisotropic physical properties. Radiative transfer theory, on the other hand, deals with the propagation of intensities. It is based on phenomenological and heuristic observations of the transport characteristics of intensities. The basic differential equation is called the transport equation and is equivalent to Boltzmann's equation in the kinetic theory of gases and in neutron transport theory. To determine the light propagation in tissue by the transport equation, only the absorption coefficient μ_a, scattering coefficient μ_s, and the scattering phase function $p(\theta)$ are needed. These parameters can be measured experimentally. The transport theory has been found very useful in tissue optics studies. The following is a detailed description of the transport theory and the Monte Carlo simulation method for numerical solutions of the transport equation. The definitions and experimental measurements of tissue optical transport parameters are also discussed.

3.1   LIGHT TRANSPORT PARAMETERS

The absorption coefficient μ_a describes the probability of light absorption in tissue. It is in units of cm^{-1}, and its inverse value $1/\mu_a$ indicates the mean free path between absorption events. The scattering coefficient μ_s describes the probability of light scattering in tissue. It is also in units of cm^{-1}, and its inverse value $1/\mu_s$ indicates the mean free path between scattering events. The anisotropy g indicates the angular deflection of a photon's trajectory caused by a scattering event. The anisotropy equals the expectation value, $<\cos \theta>$, where θ is the photon deflection angle. As shown in Figure 3.2, the scattered photons have uniform distributions along the azimuthal angle ψ. The effectiveness of light scattering is described by the product $\mu_s(1 - g)$. A g of zero, corresponding to isotropic scattering that scatters light randomly in all directions, implies that the scattering is 100% effective. If g equals 0.9, then a photon must be scattered 10 times to achieve the randomization equivalent to a single isotropic scattering event. If g equals 1, it corresponds to total forward

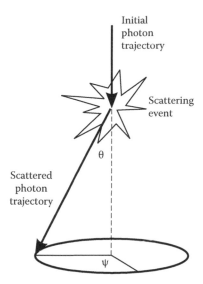

FIGURE 3.2   Deflection of a photon by a scattering event. The angle of deflection, θ, and the azimuthal angle, ψ, are indicated.

scattering. The effective (or reduced) scattering coefficient, $\mu_s(1 - g)$, or simply μ_s', is in units of cm^{-1} too, and its inverse value $1/\mu_s'$ indicates the mean free path before a photon's trajectory becomes randomized.

3.2   TRANSPORT THEORY

A basic quantity in the transport equation is the specific intensity or radiance $L(r,s)$ which denotes the average power flux density within a unit solid angle along direction s, at position r, and has the unit W/cm^2/sr. The fluence rate (or light intensity) $\Phi(r)$ (its quantity is the fluence or radiant flux rate per unit area) is related to $L(r,s)$ by

$$\Phi(r) = \int_{4\pi} L(r,s)\,d\omega \tag{3.1}$$

where $d\omega$ is the differential solid angle.

If a specific intensity $L(r,s)$ is independent of the direction s, then the radiation is said to be "isotropic." If the specific intensity radiated from a surface da is isotropic, then the power P radiated from this surface da in the direction s is given by

$$P\ (\text{W sr}^{-1}) = (Lda)\cos\theta = P_0\cos\theta \tag{3.2}$$

where θ is the angle between the direction s and the normal to the surface da. This relationship (Equation 3.2) is called Lambert's cosine law.

Although the transport theory ignores the wave properties of the radiation, the radiance can be proven to be the sum of all Poynting vectors whose tips are located within a solid angle in the direction s and related to the mutual coherence function used in the analytical theory through a Fourier transform. This also means that even though transport theory was developed on the basis of the addition of powers, it contains information about the correlation of the fields (Ishimaru 1978).

The transport equation reads:

$$(s \cdot \nabla)L(r,s) = -(\mu_a + \mu_s)L(r,s) + \mu_s \int_{4\pi} p(s,s')L(r,s')d\omega' + \varepsilon(r,s) \qquad (3.3)$$

where μ_a (cm^{-1}) and μ_s (cm^{-1}) are the absorption and scattering coefficients, respectively, $p(s,s')$ is the *phase function*[*] representing the probability that a photon is scattered from direction s into direction s'. It depends only on the angle θ between s and s' assuming that the scatters are randomly distributed over the tissue volume. The Henyey–Greenstein phase function is a good approximation for biological tissues (Jacques et al. 1987):

$$p(\theta) = \frac{1}{4\pi} \frac{1 - g^2}{(1 + g^2 - 2g\cos\theta)^{3/2}} \qquad (3.4)$$

where g is the scattering anisotropy. The integration in Equation (3.3) is over all 4π steradians of solid angle in a spherical coordinate system. Equation (3.3) is the result of energy conservation. The left side of the equation denotes the radiance change rate in the s direction. The first term of the right side denotes the losses in $L(r,s)$ per unit of length in the s direction due to absorption and scattering. The second term denotes the gain in $L(r,s)$ per unit of length in the s direction due to scattering from all other direction s'. The radiance $L(r,s)$ may also increase due to light emission (e.g., fluorescence emission) from within the volume ds. The $\varepsilon(r,s)$ denotes the power radiation per unit volume per unit solid angle in the s direction.

Equation (3.3) is a local integro-differential equation. No general accurate analytical solutions are available for a turbid biological tissue with strong forward scattering properties. The most successful approximation method is the diffusion theory. However, it is a good approximation only when scattering dominates over absorption and deals with situations far away from the light source and boundaries. For details on diffusion theory, see references (Ishimaru 1978; Jacques and Prahl 1987; Patterson et al. 1991a, b; Wang and Wu 2007). However, Monte Carlo simulation methods have been developed to obtain rigorous numerical solutions of the transport equation.

---

[*] The term *phase function* has its origin in astronomy where it refers to lunar phases. It has no relation to the phase of a wave. (See Ishimaru 1978.)

3.3 MONTE CARLO SIMULATION OF LIGHT PROPAGATION IN BIOLOGICAL TISSUE

In Monte Carlo simulation, the step-by-step trajectories of individual photons in tissue are calculated by computer. Specifically, light is treated as noninteracting photon bundles, and tissue is stochastically identified by probability density functions for light attenuation, $F(s)$, and direction of light scattering, $G(\psi)$ and $H(\theta)$:

$$F(s) = \mu_t \exp(-\mu_t s) \tag{3.5}$$

$$G(\psi) = 1/2\pi \tag{3.6}$$

$$H(\theta) = \frac{1 - g^2}{2(1 + g^2 - 2g \cos\theta)^{3/2}} \tag{3.7}$$

In Equation (3.5), s is the photon bundle pathlength between attenuation events, and $\mu_t\,(= \mu_a + \mu_s)$ is the total attenuation coefficient. In Equations (3.6) and (3.7), ψ and θ identify the new direction of the scattered photon in a spherical geometry. Here, θ is the deflection angle, while ψ is the azimuthal angle. A computer-generated random number is used to determine the probability of an event (F, G, H) occurring, and the probabilistic parameter (s, ψ, θ) is calculated. The photon bundle is propagated through the tissue by absorbing a fraction μ_a/μ_t of its "weight" after the bundle has traveled the pathlength s. The remaining weight is scattered in the new direction specified by ψ and θ, measured from the previous photon direction. Refractive index mismatched boundary conditions are handled by calculation of the Fresnel reflection coefficients, assuming unpolarized light:

$$R(\theta_i) = \frac{1}{2}\left[\frac{\sin^2(\theta_i - \theta_t)}{\sin^2(\theta_i + \theta_t)} + \frac{\tan^2(\theta_i - \theta_t)}{\tan^2(\theta_i + \theta_t)}\right] \tag{3.8}$$

where θ_i is the incident angle of the photon onto the boundary ($\theta_i = 0$ implies orthogonal incidence), θ_t is the deflection angle of the transmitted photon and is related to θ_i by Snell's law:

$$n_i \sin\theta_i = n_t \sin\theta_t \tag{3.9}$$

with n_i and n_t being the refractive indices of the media. The photons are multiply scattered by tissue and the phase and polarization are quickly randomized, and play little role in the energy transport.

The simulation program records the accumulated light power density, Q, in W/cm^3 that is deposited in a local tissue volume. The local light fluence Φ, in W/cm^2 is calculated using the local absorption coefficient μ_a, in cm^{-1}:

$$\Phi = Q/\mu_a \tag{3.10}$$

Φ represents the light distribution in tissue. The program also records where and in which direction the photons escape the air–tissue interfaces. Therefore, the local diffuse reflectance and transmittance as well as the total reflectance and transmittance can be obtained.

Tissue autofluorescence measurements can be modeled with the following procedure:

1. Excitation light distribution. The distribution of excitation light within the tissue must be specified first. It is calculated using the Monte Carlo simulation, and is denoted as $\Phi(\lambda_{ex}, r, z, \theta)$ in W/cm². λ_{ex} is the excitation wavelength, while r, z, θ represent a local position in cylindrical coordinates.

2. Intrinsic fluorescence coefficient, $\beta(\lambda_{ex}, \lambda_{em}, z)$. The intrinsic fluorescence coefficient β is defined as the product of the absorption coefficient due to the fluorophore, μ_{afl} (cm⁻¹), and the quantum yield Y (dimensionless) of fluorescence emission (Keijzer et al. 1989). Biological tissues usually have a layered structure. In the same layer, β is constant, therefore, it can be denoted as a function of z, $\beta(\lambda_{ex}, \lambda_{em}, z)$. λ_{em} is the wavelength of emitted fluorescence light. The product $\Phi\beta$ yields the density of fluorescence source in W/cm³. Using a microspectrophotometer system and frozen tissue sections, one can measure the relative β distribution in the tissue (Zeng et al. 1993, 1997).

3. Escape function $E(\lambda_{em}, r, z)$. Once a fluorophore emits a fluorescent photon, that photon must successfully reach the surface and escape to be observed. The escape function $E(\lambda_{em}, r, z)$ is the surface distribution as a function of radial position (r) of escaping photons from a point source of fluorescence at depth z and radial position $r = 0$ within a tissue of thickness D. It also can be calculated by Monte Carlo simulations. The units of E are (W/cm² escape at surface)/(W of source point within tissue) which equals cm⁻². Simulations are conducted for a series of depth (z) inside the tissue, using the optical properties for the emission wavelengths of interest.

4. Observed fluorescence, $F(\lambda, r)$. The observed flux of escaping fluorescence F in W/cm² at the tissue surface is computed by the following convolution (Keijzer et al. 1989):

$$F(\lambda_{ex},\lambda_{em},r) = \int_0^D \int_0^{2\pi} \int_0^{\infty} \Phi(\lambda_{ex},r',z',\theta')\beta(\lambda_{ex},\lambda_{em},z') \times$$
$$E(\lambda_{em},\sqrt{r^2 + r'^2 - 2rr'\cos\theta'},z')r'dr'd\theta'dz' \tag{3.11}$$

The convolution in Equation (3.11) can be implemented numerically using discrete values for Φ and E that were generated by the Monte Carlo simulations, and the experimentally determined β.

Tissue Raman measurements can also be modeled in a similar fashion by replacing β with a Raman scattering efficiency quantity.

For details of the implementation of the Monte Carlo simulation, see (Prahl et al. 1989; Wang et al. 1995, 1997). A simulation program of "Monte Carlo Modeling of Light Transport in Multi-layered Tissue in Standard C (MCML)" is available in the public domain (http://labs.seas.wustl.edu/bme/Wang/mc.html or http://omlc.ogi.edu/software/mc/). The MCML program can be used to calculate the fluence distribution inside layered tissues and the diffuse reflectance for a normal incident light beam on the tissue surface. To calculate the fluorescence escape function, the MCML program can be modified to simulate the light propagation process for a buried isotropic point source inside the tissue. Example applications of modeling tissue reflectance spectral measurements and fluorescence spectral measurements can be found in (Zeng et al. 1997, Chen et al. 2007).

3.4 MEASUREMENT OF TISSUE TRANSPORT PARAMETERS

Tissue optical properties (transport parameters) are not only the prerequisite for theoretical modeling, but can also provide useful diagnostic information if derived from *in vivo* measurements. In the early days, measurements are mostly carried out with excised (*ex vivo*) tissue samples, while recently *in vivo* method development is intensively studied.

A host of investigators have developed various techniques and methods to measure the three transport parameters (μ_a, μ_s, g) of various tissues *ex vivo* (Cheong et al. 1990). In summary, any three of the following four measurements will be sufficient to determine the three parameters: (1) total (or diffuse) transmission for collimated or diffuse irradiance, (2) total (or diffuse) reflection for collimated or diffuse irradiance, (3) unscattered (collimated) transmission for collimated irradiation, and (4) angular distribution of emitted light from an irradiated sample. The data processing methods for determining the three parameters can be classified into the following two categories:

(a) *Direct methods,* which do not depend on any specific transport model to obtain the optical parameter from measurements. For example, μ_t ($= \mu_a + \mu_s$) can be derived directly from the above measurement (4) using Beer's Law:

$$T_c = \exp(-\mu_t t) \qquad (3.12)$$

where T_c is the unscattered transmittance, and t the thickness of the tissue slab.

(b) *Iterative indirect methods,* which use complicated solutions to the transport equation. Examples are diffusion theory (Jacques and Prahl 1987), adding-doubling model (Prahl et al. 1993), and Monte Carlo simulation (Peter et al. 1990). Iterative algorithms may be computing intensive, but generate more accurate parameters.

Table 3.1 summarizes these efforts in terms of measurement type, transport parameters involved, and data analysis methods for deriving the related optical parameters.

TABLE 3.1

Methods for Deriving Transport Parameters from *Ex Vivo* Tissue Measurements

Measurements	Transport Parameters	Data Analyses
Unscattered transmission of thin tissue slab ($\sim$ 100 µm)	$\mu_t\ (= \mu_a + \mu_s)$	Direct method (Beer's law)
Angular distribution of scattered light by thin tissue slab (< 100 µm)	g	Direct method (Henyey-Greenstein phase function)
Total reflectance from semi-infinite media (very thick tissue)	$\dfrac{\mu_s(1-g)}{\mu_a}$	Indirect method (Monte Carlo simulation or diffusion theory)
Total reflectance and total transmission of thick tissue (200–2000 µm)	μ_a and $\mu_s'\ (= \mu_s\,(1-g))$	Indirect method (diffusion theory, Monte Carlo simulation, and adding-doubling model)

Figure 3.3 shows example optical properties of normal bronchial tissues (epithelium, submucosa, cartilage) and tumor tissue measured from *ex vivo* tissue samples (Qu et al. 1994). The three measurements used in this study were unscattered transmission, total reflectance, and total transmission, while the inverse adding-doubling solution was used to analyze the data and derive optical parameters. A comprehensive list of tissue optical parameters compiled from existing literature can be found in (Mobley and Vo-Dinh 2003). The scattering anisotropy g values of biological tissues are usually large (>0.7), indicating that light scattering by biological tissues is strongly forward directed. In the visible to near IR wavelength range, the absorption coefficient μ_a and scattering coefficient μ_s usually decrease with increasing wavelengths. This creates a so-called "imaging and therapeutic window" from 600 nm to 1400 nm where light can penetrate deeper into the tissue. Above 1400 nm, the very rich water content in tissue exhibits strong absorption to IR light.

More recently, *in vivo* measurements such as spatially resolved, time-resolved, and frequency domain reflectance are being exploited to determine the tissue transport parameters non-invasively (Wilson and Jacques 1990 and references within, Fishkin et al. 1997; Shah et al. 2001). Other noninvasive methods include pulsed photothermal radiometry (Anderson et al. 1989; Prahl et al. 1992), oblique-incidence reflectometry (Wang and Jacques 1995; Garcia-Uribe et al. 2009), structured illumination imaging (Cuccia et al. 2005), and reflectance confocal microscopy/OCT (optical coherence tomography) (Samatham et al. 2008). However, most of these methods return the average optical parameters over a large tissue volume containing multiple tissue layers. The reflectance confocal microscopy/OCT methods are capable of deriving scattering properties (both μ_s and g) of individual tissue layers. Application of these methods to endoscopy is often a challenging task. To accurately derive the optical parameters from these *in vivo* measurements, more theoretical studies on solving the inverse problem of multi-layered media are also required.

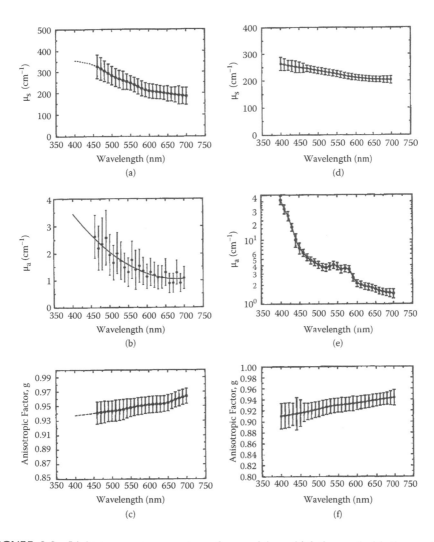

FIGURE 3.3 Light transport parameters of normal bronchial tissues (epithelium, submucosa, cartilage) and tumor tissue measured from *ex vivo* tissue samples. (a) scattering coefficient μ_s of the epithelium, (b) absorption coefficient μ_a of the epithelium, (c) scattering anisotropy **g** of the epithelium, (d) μ_s of the submucosa, (e) μ_a of the submucosa, and (f) **g** of the submucosa. (Reproduced with permission from Qu, J. et al. 1994. *Appl. Opt.* 33: 7397–7405.)   *Continued*

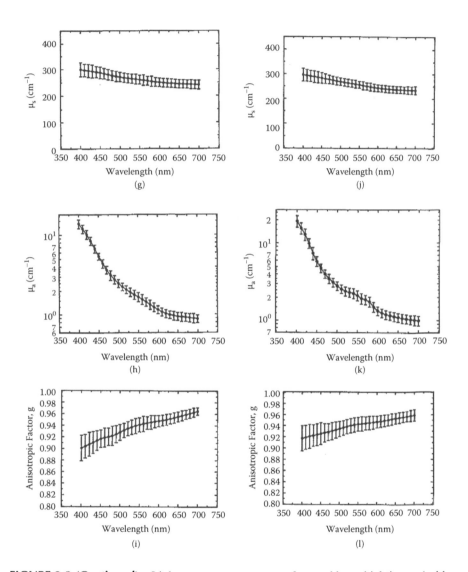

FIGURE 3.3 (Continued)   Light transport parameters of normal bronchial tissues (epithelium, submucosa, cartilage) and tumor tissue measured from *ex vivo* tissue samples. (g) μ_s of the cartilage, (h) μ_a of the cartilage, (i) *g* of the cartilage, (j) μ_s of the tumor, (k) μ_a of the tumor, and (l) *g* of the tumor. (Reproduced with permission from Qu, J. et al. 1994. *Appl. Opt.* 33: 7397–7405.)

REFERENCES

Anderson, R.R., Beck, H., Bruggemann, U., Farinelli, W., Jacques, S.L., and Parrish, J.A. 1989. Pulsed photothermal radiometry in turbid media: Internal reflection of backscattered radiation strongly influences optical dosimetry. *Appl. Opt.* 28: 2256–2262.

Chen, R., Huang, Z., Lui, H., Hamzavi, I., McLean, D.I., Xie, S., and Zeng, H. 2007. Monte Carlo simulation of cutaneous reflectance and fluorescence measurements—the effect of melanin contents and localization. *J. Photochem. Photobiol. B: Biol.* 86: 219–226.

Cheong, W.F., Prahl, S.A., and Welch, A.J. 1990. A review of the optical properties of biological tissues. *IEEE J. Quantum Electronics* 26: 2166–2184.

Cuccia, D.J., Bevilacqua, F., Durkin, A.J., and Tromberg, B.J. 2005. Modulated imaging: Quantitative analysis and tomography of turbid media in the spatial-frequency domain. *Opt. Lett.* 30: 1354–1356.

Fishkin, J.B., Coquoz, O., Anderson, E.R., Brenner, M., Tromberg, B.J. 1997. Frequency-domain photon migration measurements of normal and malignant tissue optical properties in a human subject. *Appl. Opt.* 36: 10–20.

Garcia-Uribe, A., Balareddy, K.C., Zou, J., Wojcik, A.K., Wang, K.K., and Wang, L.V. 2009. Micromachined side-viewing optical sensor probe for detection of esophageal cancers. *Sensor. Actuat. A: Physical* 150: 144–150.

Ishimaru, A. 1978. *Wave Propagation and Scattering in Random Media. Vol. 1.* New York: Academic.

Jacques, S.L. and Prahl, S.A. 1987. Modeling optical and thermal distributions in tissue during laser irradiation. *Lasers Surg. Med.* 6: 494–503.

Keijzer, M., Richards-Kortum, R., Jacques, S.L., and Feld, M.S. 1989. Fluorescence spectroscopy of turbid media: Autofluorescence of the human aorta. *Appl. Opt.* 28: 4286–4292.

Mobley, J. and Vo-Dinh, T. 2003. Optical properties of tissue. In *Biomedical Photonics Handbook*, ed. T. Vo-Dinh. Boca Raton: CRC Press, pp. 2-1–2-75.

Patterson, M.S., Wilson, B.C., and Wyman, D.R. 1991a. The propagation of optical radiation in tissue. I: Models of radiation transport and their application. *Laser Med. Sci.* 6: 155–168.

Patterson, M.S., Wilson, B.C., and Wyman, D.R. 1991b. The propagation of optical radiation in tissue. II: Optical properties of tissues and resulting fluence distributions. *Laser Med. Sci.* 6: 379–390.

Peter, V.G., Wyman, D.R., Patterson, M.S., and Frank, G.L. 1990. Optical properties of normal and diseased human breast tissues in the visible and near infrared. *Phys. Med. Biol.* 35: 1317–1334.

Prahl, S.A., Keijzer, M., Jacques, S.L., and Welch, A.J. 1989. A Monte Carlo model of light propagation in tissue. In *Dosimetry of Laser Radiation in Medicine and Biology, SPIE Series IS* 5: 102–111.

Prahl, S.A., Vitkin, I.A., Bruggemann, U., Wilson, B.C., and Anderson, R.R. 1992. Determination of optical properties of turbid media using pulsed photothermal radiometry. *Phys. Med. Biol.* 37: 1203–1217.

Prahl, S.A., van Gemert, M.J.C., and Welch, A.J. 1993. Determining the optical properties of turbid media using the adding-doubling method. *Appl. Opt.* 32: 559–568.

Qu, J., MacAulay, C., Lam, S., and Palcic, B. 1994. Optical properties of normal and carcinomatous bronchial tissue. *Appl. Opt.* 33: 7397–7405.

Samatham, R., Jacques, S.L., and Campagnola, P. 2008. Optical properties of mutant versus wild-type mouse skin measured by reflectance-mode confocal scanning laser microscopy (rCSLM). *J. Biomed. Opt.* 13: 041309.

Shah, N., Cerussi, A., Eker, C., Espinoza, J., Butler, J., Fishkin, J., Hornung, R., and Tromberg, B. 2001. Non-Invasive Functional Optical Spectroscopy of Human Breast Tissue. *Proc. Natl. Acad. Sci.* 98: 4420–4425.

Wang, L.V. and Jacques, S.L. 1995. Use of a laser beam with an oblique angle of incidence to measure the reduced scattering coefficient of a turbid medium. *Appl. Opt.* 34: 2362–2366.

Wang, L.-H., Jacques, S.L., and Zheng, L.-Q. 1995. MCML—Monte Carlo modeling of photon transport in multi-layered tissues. *Comput. Methods Programs Biomed.* 47: 131–146.

Wang, L.-H., Jacques, S.L., and Zheng, L.-Q. 1997. CONV—Convolution for responses to a finite diameter photon beam incident on multi-layered tissues. *Comput. Methods Programs Biomed.* 54: 141–150.

Wang, L.V. and Wu, H.-I. 2007. *Biomedical Optics: Principles and Imaging.* Hoboken, NJ: John Wiley & Sons.

Wilson, B.C. and Jacques, S.L. 1990. Optical reflectance and transmittance of tissue: Principles and applications. *IEEE J. Quantum Electronics* 26: 2186–2198.

Zeng, H., MacAulay, C., McLean, D.I., and Palcic, B. 1993. Novel microspectrophotometer and its biomedical application. *Opt. Eng.* 32: 1809–1813.

Zeng, H., MacAulay, C., McLean, D.I., and Palcic, B. 1997 Reconstruction of *in vivo* skin autofluorescence spectrum from microscopic properties by Monte Carlo simulation. *J. Photochem. Photobiol. B: Biol.* 38: 234–240.

Section II

Endoscopic Field
Imaging Modalities

4 White Light Endoscopy

Haishan Zeng
British Columbia Cancer Agency Research Centre,
Vancouver, British Columbia, Canada

CONTENTS

White light endoscopy is the basic imaging modality in any endoscopes. Through the insertion of an endoscope, this imaging modality enables a physician to visualize the internal organ, its color appearance, surface morphology, and texture. It covers a field of view of about a centimeter to several centimeters across. It not only serves as a diagnostic tool, but also provides visual guidance for various interventions such as sampling of liquid, cellular, and suspicious tissues (biopsy) and performing therapies, for example, electrocautery, laser ablation, and PDT (photodynamic therapy). An image is usually formed under white light illumination although varieties are introduced by altering the illumination, introducing exogenous dyes, or increasing magnification in order to extend its diagnostic capabilities. The configuration of ordinary white light endoscopy systems have been discussed in Chapter 2. This chapter will focus on the origins of signal, advantages and shortfalls, and varieties of this imaging modality.

4.1 ORIGINS OF THE SIGNALS

The origins of signals in conventional white light endoscopy and its varieties are all from light reflection. A more general terminology for white light endoscopy and its varieties should be "reflectance imaging" or "reflectance endoscopy." As shown in Chapter 1, Figure 1.2, there are two types of light reflections: specular reflection (also called *mirror reflection*) and diffuse reflection. Diffuse reflection dominates the signal in white light endoscopy imaging, however, bright specular reflection spots are occasionally visible, especially when the tissue surface is wet. Specular reflection is the direct surface reflection of the illumination light by the air–tissue interface. The signal strength is determined by the refractive index differences between the air and the tissue and can be calculated using the Fresnel reflection coefficient as described in Chapter 3, Equation (3.8). The reflected light rays forms the same angle to the surface normal as the incident light, but at the opposite side and direction (Figure 1.2). Therefore, the specular reflection can be easily missed from being collected by the endoscope imaging optics due to the surface roughness or the relative angle of the endoscope to the tissue surface. In white light color photography of skin lesions, polarization optics are often used to either enhance the specular reflection collection for surface texture assessment or to remove the specular reflection collection for better assessment of the internal tissues. However, it is technically inconvenient to implement polarization optics in endoscopy.

Diffuse reflection is originated from the portion of the refracted incident light that has been absorbed and scattered (often multiple times) by the tissue and reemerged out of the air–tissue interface (Figure 1.2). The amount of diffuse reflection is determined by the absorption and scattering properties of the tissue. The stronger the absorption, the less the diffuse reflection; the stronger the scattering, the larger the diffuse reflection. For a semi-infinite tissue-like media, the diffuse reflectance, R is a function of the ratio of the reduced scattering coefficient, $\mu_s(1 - g)$, to the absorption coefficient, μ_a:

$$R = F[\mu_s(1 - g)/\mu_a] \tag{4.1}$$

In the visible wavelength range as pertained to white light endoscopy imaging, the dominant chromophores that absorbs light in tissues of internal organs are hemoglobin and oxyhemoglobin. In the infrared (IR), the light absorption is dominated by water, while in the ultraviolet (UV) by proteins and DNA. Figure 4.1 shows the absorption spectra for hemoglobin and oxyhemoglobin. The scattering properties of tissue are dominated by the lipid–water interfaces presented by membranes in cells (especially membranes in mitochondria), by nuclei, and by protein fibers such as collagen and elastin.

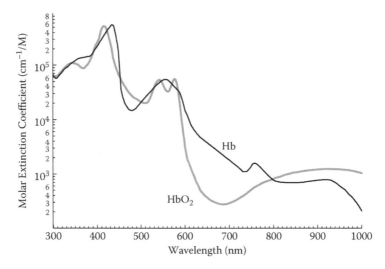

FIGURE 4.1   Absorption spectra of hemoglobin and oxyhemoglobin.

From the above analysis, reflectance imaging (white light endoscopy and its varieties) contains information about tissue surface roughness and textures as well as volume tissue scattering and absorption properties that are affected by tissue micromorphological structures and physiological status such as blood supply and oxygenation. On the other hand, surface morphology can also be perceived by reflectance imaging because both the illumination intensity and the light collection efficiencies for both specular reflection and diffuse reflection for a specific location on the tissue surface (corresponding to a particular pixel in the image) is inversely proportional to the square of the distance between the surface location and the endoscope tip. Reflectance imaging also allows a physician to perceive the color of the tissue field under observation. Specular reflection is very weakly dependent on the wavelength of light and dominates the surface roughness and texture perception. Diffuse reflectance is strongly wavelength dependent and dominates the color perception. Surface morphology affects the collection efficiency of both diffuse reflection and specular reflection.

4.2 COLOR PERCEPTION

The basic and "must have" reflectance endoscopy imaging modality is the white light color imaging that is provided by every endoscopy system. White light color imaging can help with diagnosis and also allows a physician to conveniently perform interventions. Visible white light (400–700 nm) is used for illumination, simultaneously or sequentially. The reflectance color image is either captured by a CCD chip located at the distal tip of the endoscope or transmitted by a coherent fiber bundle to the proximal end of the endoscope, where it can be observed directly by the physician's eye or captured by a CCD color camera. The captured imaging videos are displayed on either a computer monitor or an analogue monitor for viewing by the physician and other personnel working in the endoscopy suite. Using color imaging is not only more pleasing, but it also enables one to receive more visual information. While we can perceive only a few dozen gray levels, we have the ability to distinguish between thousands of colors.

The human eye senses light using a combination of rod and cone cells. Rod cells are better for low-light vision, but can only sense the intensity of light. Cone cells are for discerning color, they function best in bright light. Three types of cone cells exist in the eye, each being more sensitive to either short (S), medium (M), or long (L) wavelength light. Figure 4.2 illustrates the relative sensitivity of each type of cone cell for the entire visible spectrum from 400 nm to 700 nm. These sensitivity curves are denoted as $S_1(\lambda)$, $S_2(\lambda)$, and $S_3(\lambda)$, respectively. Note how each type of cone does not just sense one color, but instead has varying degrees of sensitivity across a broad range of wavelengths. There are significant overlap between $S_1(\lambda)$ and $S_2(\lambda)$. The set of signals possible at all three cone cells describes the range of colors we can see with our eyes.

Based on the trichromatic theory of color vision (Young 1802; Jain 1989), light with a spectral profile, $C(\lambda)$, will produce a color sensation that can be described by spectral responses as

$$\alpha_i(C) = \int S_i(\lambda) \cdot C(\lambda) d\lambda, \qquad i = 1, 2, 3 \tag{4.1}$$

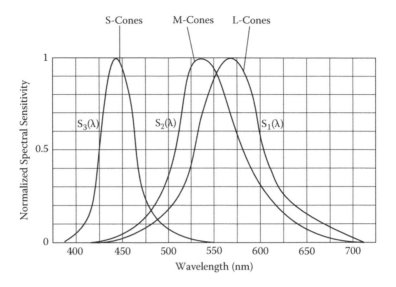

FIGURE 4.2   Normalized spectral sensitivity of human vision cone cells, S, M, and L.

where α_1, α_2, and α_3 represents the three primary color components: red (R), green (G), and blue (B), respectively. This equation describes the color perception quantitatively. If $C_1(\lambda)$ and $C_2(\lambda)$ are two spectral distributions that produce responses $\alpha_i(C_1)$ and $\alpha_i(C_2)$ such that

$$\alpha_i(C_1) = \alpha_i(C_2), \qquad i = 1, 2, 3 \tag{4.2}$$

then the colors C_1 and C_2 are perceived to be identical. Hence, two colors that look identical could have different spectral distributions. When α_1, α_2, and α_3 all reach the maximum values, it represents "white" color, while $\alpha_1 = \alpha_2 = \alpha_3 = 0$ represents black. The properties of different colors can be well described by three perceptual attributes: brightness (or lightness), hue, and saturation. Figure 4.3 shows a perceptual representation of the color space. Brightness (W^*) represents the perceived luminance. It varies along the vertical axis. The hue (H) of a color is determined by its dominant wavelength. When we call a color "red," we are referring to its hue. The hue varies along the circumference. The saturation (S) of a color is a measurement of its purity or how much it has been diluted by white. The saturation varies along the radius. In Figure 4.3, for a fixed brightness W^*, the symbols R, Y, G, and B show the relative locations of the red, yellow, green, and blue spectral colors.

It has to be noted that commercial color CCD cameras used in a white light endoscopy system often have different spectral sensitivity from that of human version. Figure 4.4 shows the normalized spectral sensitivity of a commercial CCD color camera that uses a Bayer filter mosaic to obtain the primary color, R, G, B images. An example white light endoscopy image taken from a colonic polyp using such a camera attached to a fiber optic colonoscope is shown in Figure 5.4(B). In Figure 4.2, the M and L curves are quite close to each other, while in Figure 4.4, the

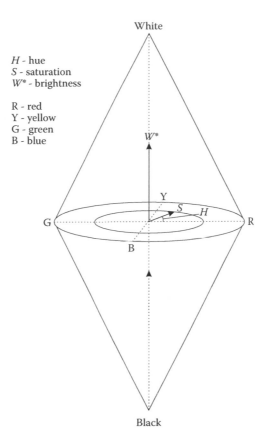

FIGURE 4.3 Perceptual representation of the color space. The brightness W^* varies along the vertical axis, hue H varies along the circumference, and the saturation S varies along the radius.

three spectral curves are quite evenly separated. Another factor that affects endoscopy color perception is the illumination. Various lamps of different spectral profiles are used in different endoscopy systems, including tungsten lamp, Xenon arc lamp, metal halide lamp, and even light emitting diodes (LEDs) (Yanai et al. 2004). Color calibration and white balance adjustment are often carried out to achieve consistent color perception within the same endoscopy system. However, these calibrations cannot completely eliminate the differences between different endoscopy systems because the individual color sensations, α_i ($i = 1, 2, 3$), have complicated relationships with the system spectral responses as described in Equation (4.1).

4.3 ADVANTAGES AND LIMITATIONS

The advantages of conventional white light endoscopy imaging is that it facilitates color vision of the internal surfaces of internal organs. It enables a visual inspection type of diagnosis and guides interventions. However, it is basically an optical device

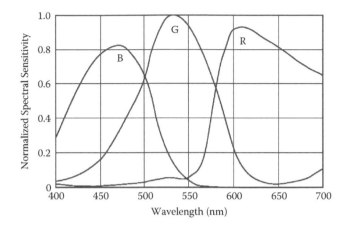

FIGURE 4.4   Normalized spectral sensitivity of a commercial CCD color camera.

optimized for human vision. It is not designed to realize the full potential of reflectance imaging for disease diagnosis. Its sensitivity for the detection of many early lesions is very low. For example, the diagnostic sensitivity of white light endoscopy for early lung cancer is only ~9% (Lam et al. 1998). This is because these lesions are only a few cell layers thick and up to a few millimeters in surface diameter, producing insufficient changes on color or surface morphology to make them visible under white light endoscopy.

A natural next step to improve the diagnostic performance of a white light endoscope is to illuminate or capture reflectance images at much narrow wavelength bands rather than the very broad R, G, B bands optimized for human vision. These narrow band images will in principle allow us to derive more detailed information about the tissue micro-morphology and physiology such as vasculature structures and tissue oxygenation.

4.4   NARROW BAND IMAGING

In narrow band imaging (NBI) mode, the illumination is restricted to two narrow bands centered at 415 nm and 540 nm with a bandwidth of about 30 nm. Corresponding reflectance images at these two narrow bands are acquired by a video endoscope in commercial devices from Olympus Medical Systems (Tokyo, Japan). In principle, the same imaging modalities can be implemented on fiber endoscope systems as well. The 415 nm wavelength is aligned with the strongest hemoglobin absorption peak as shown in Figure 4.1. This wavelength penetrates only the superficial epithelial tissue layers and is strongly absorbed by the capillary vessels. The 540 nm wavelength is at another strong absorption band of hemoglobin, and it can penetrate deeper below the capillaries and get absorbed more by the larger blood vessels such as the veins. NBI modality provides enhanced endoscopy visualization of superficial neoplastic lesions and the microvascular architectures. It can also be combined with a magnifying endoscope through either digital or optical zoom to further enhance the image texture details.

The first clinical use of the NBI system was reported by (Sano et al. 2001) for gastrointestinal diagnosis. Their promising observations resulted in a subsequent pilot colorectal study in which the NBI system demonstrated better vascular pattern visualization than conventional white light colonoscopy in the diagnosis of colorectal polyps (Machida et al. 2004). NBI clinical uses are now expanded to the diagnosis of premalignant and malignant lesions of the hypo-pharynx, esophagus, and stomach. Detailed reviews of the NBI technology and its clinical applications can be found in (Emura et al. 2008; Muto et al. 2009). Example NBI images can be found in Chapter 13, Figures 13.3 and 13.4.

4.5 IMAGE CONTRAST ENHANCEMENT BY EXOGENOUS AGENTS

Using the administration of an exogenous agent is a very simple method to improve image contrast and therefore the diagnostic capability of white light endoscopy. Dye solutions are applied to the mucosa of internal organs first; conventional white light endoscopy is then used to image the mucosa. The imaging can also be further enhanced by using a magnifying endoscope. This endoscopy technique is called *chromoendoscopy*, or vital staining, and contrast endoscopy. It often results in better recognition of subtle mucosal changes that are invisible by direct white light endoscopy. The dyes can be administrated through the instrument channel of an endoscope by using a special catheter with fine holes at the distal tip to enable a uniform spray. The exogenous agent used in chromoendoscopy can be classified into three different types.

1. Contrast dyes. It simply coats the mucosal surface and neither reacts with nor gets absorbed by the mucosa. It helps highlight the surface textures of the mucosa. Indigo carmine is an example of this type.
2. Absorptive dyes. It gets absorbed by different cells and microstructures to different degrees. This brings imaging contrast between different cell types. An example of this type of dye is methylene blue, widely used in colonoscopy.
3. Reactive dyes. It reacts with certain mucosa components and changes the color. Acetic acid is an example of this type.

A detailed review of chromoendoscopy can be found in (Kiesslich and Neurath 2006). Example clinical images taken with this technology are shown in Chapter 13, Figures 13.1 and 13.2.

4.6 NEW DEVELOPMENT—SPECTRAL IMAGING

The goal of spectral imaging is to obtain a spectrum for each pixel of the image. This is the best way to fully make use of the reflectance information a white light endoscopy (or more precisely a reflectance endoscopy) could possibly provide. To obtain a spectral image, a few tens of narrow band images have to be acquired. Spectral imaging for biomedical applications has been successfully implemented on microscopy settings and *in vivo* for skin diagnosis. For example, Stamatas et al. 2006 used spectral imaging to obtain *in vivo* 2-D maps of various skin chromophores including oxy- and deoxyhemoglobins. They used 18 narrow band filters (10 nm bandwidth) to obtain images in the 400–970 nm range. A phase correction algorithm was used to align the individual images at different wavebands to fight motion artifacts. Spectral analysis is used to derive chromophore concentrations. So far, the most advanced endoscopic spectral imaging system was reported by the Farkas group (Lindsley et al. 2004; Chung et al. 2006). Their system acquires parallel and perpendicular polarized images at 32 evenly spaced bands from 380–690 nm in 0.25 seconds. A monochromator or AOTF (acousto-optic tunable filter) based tunable light source and a high speed CCD camera was used for image acquisition. A 2.0 mm size catheter consisting of an illumination fiber and two imaging fiber bundles was passed through the instrument channel of an endoscope to perform the spectral imaging. In a clinical test, they encountered problems from specular reflection interference and motion artifacts due to movement from either the endoscope or patient. The spectral imaging has to be retaken if a motion artifact occurs.

For spectral imaging to be practically useful in endoscopy, a complete image set (not just a single band image within the set) has to be acquired at the video rate (30 image sets/second) to fight the motion artifact. This imposes a significant technical challenge and has not been achieved yet. Real-time data processing is also a difficult task due to the large amount of data involved. Currently, many groups, including ourselves, are working on overcoming these technical obstacles. If successful, this new technology will greatly improve the diagnostic capabilities of reflectance endoscopy.

REFERENCES

Chung, A., Karlan, S., Lindsley, E., Wachsmann-Hogiu, S., and Farkas D.L. 2006. *In vivo* cytometry: A spectrum of possibilities, *Cytometry Part A,* 69A: 142–146.

Emura, F., Saito, Y., and Ikematsu, H. 2008. Narrow-band imaging optical chromocolonoscopy: Advantages and limitations. *World J. Gastroenterol.* 14: 4867–4872.

Jain, A.K. 1989. *Fundamentals of Digital Image Processing.* Englewood Cliffs, NJ: Prentice Hall.

Kiesslich, R. and Neurath, M.F. 2006. Chromoendoscopy and other novel imaging techniques. *Gastroenterol. Clin. N. Am.* 35: 605–619.

Lam, S., Kennedy, T., Unger, M., Miller, Y.E., Gelmont, D., Rusch, V., Gipe, B., Howard, D., LeRiche, J.C., Coldman, A., and Gazdar, A.F. 1998. Localization of bronchial intraepithelial neoplastic lesions by fluorescence bronchoscopy. *Chest* 113: 696–702.

Lindsley, E., Wachman, E.S., and Farkas, D.L. 2004. The hyperspectral imaging endoscope: A new tool for *in vivo* cancer detection, *SPIE Proc.* 5322: 75–82.

Machida, H., Sano, Y., Hamamoto, Y., Muto, M., Kozu, T., Tajiri, H. et al. 2004. Narrow-band imaging in the diagnosis of colorectal lesions: A pilot study. *Endoscopy* 36: 1094–1098.

Muto, M., Horimatsu, T., Ezoe, Y., Hori, K., Yukawa, Y., Morita, S., Miyamoto, S., and Chiba, T. 2009. Narrow-band imaging of the gastrointestinal tract. *J. Gastroenterol.* 44: 13–25.

Sano, Y., Kobayashi, M., Hamamoto, Y., Kato, S., Fu, K.I., Yoshino, T. et al. 2001. New diagnostic method based on color imaging using narrow-band imaging (NBI) system for gastrointestinal tract. *Gastrointest. Endosc.* 53: AB125.

Stamatas, G. N., Southall, M., and Kollias N. 2006. *in vivo* monitoring of cutaneous edema using spectral imaging in the visible and near infrared. *J. Invest. Dermatol.* 126: 1753–1760.

Yanai, H., Nakamura, H., and Okita, K. 2004. Development of the white light-emitting diode-illuminated digestive endoscope. *Gastrointest. Endosc.* 59: 145.

Young, T. 1802. On the theory of light and colors. *Philos. Trans. Royal Soc. London*, 92: 20–71.

5 Fluorescence Endoscopy

Haishan Zeng
British Columbia Cancer Agency Research Centre,
Vancouver, British Columbia, Canada

CONTENTS

Native tissue fluorescence, also called *tissue autofluorescence*, was observed as early as 1908 from human skin, according to Anderson and Parrish (1982). The device used is now called "Wood's lamp," which consists of a mercury discharge lamp and a UVA-transmitting, visible-absorbing filter. When using this device to observe skin autofluorescence, apparently, the eyes serve as both the *image* detector and the long pass filter. The eye is not sensitive to the UV light, but is very sensitive to the visible light. Dermatologists have since used the organism's particular fluorescence to diagnose infections such as tinea capitis, erythrasma, and some *Pseudomonas* infections. The Wood's lamp can also be used to detect porphyrins in hair, skin, or urine (Caplan 1967).

The introduction of fluorescence imaging into endoscopy diagnosis, however, was started from imaging fluorescence of exogenous agents (Profio et al. 1979). In the 1980s, with the availability of porphyrin derivatives as tumor markers, intensive research, and development of fluorescence imaging systems for improving lung cancer detection have been pursued by many groups (Hirano et al. 1989; Baumgartner et al. 1987; Potter and Mang 1984; Wagnieres et al. 1990; Anderson et al. 1987; Montan et al. 1985). The imaging is based on the strong fluorescence emission of porphyrin derivatives, such as Photofrin (Quadra Logic, Vancouver, Canada), at 630 nm and 690 nm when excited by blue light. The concentration of Photofrin in malignant tissue is higher than the adjacent normal tissues from about 3 hours on after intravenous injection. Tumor can be visualized by more intense red fluorescence than surrounding normal tissue due to Photofrin concentration differences. However, a major problem of these methods is the associated skin photosensitivity after Photofrin administration, which can last for 4 weeks or longer. Although this is not a major problem when using Photofrin for photodynamic therapy, it is clinically unacceptable for diagnostic purposes. Therefore, researchers started to lower the dose of Photofrin when used for tumor detection in order to minimize the skin photosensitivity (Lam et al. 1990). Further studies with low non-skin photosensitizing doses of Photofrin for tumor detection also found that the tissue autofluorescence exhibits significant decreases in tumor as compared to normal tissue, thus autofluorescence alone can be used for tumor detection without the administration of photosensitizer drugs (Palcic et al. 1991).

Autofluorescence endoscopy, eliminated the use of a photosensitizer and thus the associated skin photosensitivity, soon gained popularity and was continuously refined since then (Lam et al. 1993; Zeng et al. 1998; Häussinger et al. 1999; Goujon et al. 2003; Zeng et al. 2004a; Chiyo et al. 2005; Ikeda et al. 2006; Lee et al. 2007). The first clinical system (LIFE™) developed by British Columbia Cancer Agency and Xillix Technology, Inc. for lung cancer detection obtained FDA approval in 1996. Nowadays, the term *fluorescence endoscopy* often means autofluorescence endoscopy without the use of exogenous agents.

5.1   WORKING PRINCIPLES

Fluorescence endoscopy is based on imaging the autofluorescence decreases in malignant tissue as compared to surrounding normal tissue. The first FDA approved clinical system (LIFE-Lung™, Xillix Technology, Inc.) uses He-Cd laser light (442 nm) coupled into the illumination light guide of a fiber optic endoscope to illuminate and excite autofluorescence emission from bronchial tissues. Tissue autofluorescence image signals are relayed by the imaging bundle in the endoscope from the distal end to the proximal end eyepiece where a two-channel intensified CCD (ICCD) camera amplifies the image signals and captures the images. Filtering optics inside the camera separates fluorescence into two spectral bands: one with wavelengths range from 480–520 nm (green band), the other with wavelength above 625 nm (red band). They are captured by two different ICCD cameras and generate two video signals: the G video signal and the R video signal. The R and G video signals are directly overlaid on an analog monitor to generate a false color visualization of the fluorescence image. Acquiring images at video rate (30 frames/second) is critical to prevent image blurring due to involuntary movement of the organ. Normal bronchial tissue has strong green fluorescence and appears as bright green color. Malignant tissues (such as high-grade dysplasia, carcinoma *in situ*, and invasive lung cancers) have largely decreased with green fluorescence. But their red fluorescence does not decrease as much, therefore malignant tissues have an elevated R/G ratio, leading to a brownish/reddish color appearance on the display. Figure 5.1 shows a white light image and fluorescence image taken from the same location in the left main bronchus of a patient with a carcinoma *in situ* lesion. A multicenter clinical trial using this system demonstrated that the relative sensitivity of fluorescence endoscopy versus white light endoscopy was 6.3 for detection of intraepithelial neoplastic lesions and 2.71 when invasive carcinomas were also included (Lam et al. 1998).

In the above mentioned LIFE-Lung™ system, the red fluorescence imaging channel serves as a reference image to correct the green fluorescence intensity for illumination nonuniformalty and detection efficiency variations over the imaged tissue

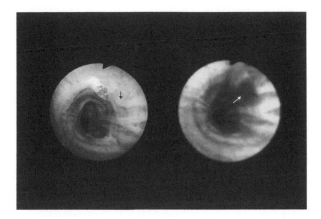

FIGURE 5.1 (See color insert)   White light image (left) and autofluorescence image (right) of a carcinoma *in situ* lesion in the left main bronchus. The white light image shows a subtle nodular lesion. The fluorescence image shows a brownish area highlighted the lesion much better than the white light image. (Reproduced with permission from Zeng, H. et al. 2004b. *Photodiagn. Photodyn. Ther.* 1: 111–122.)

area. But this reference is not ideal because the red fluorescence intensity is not constant between lesion and normal tissue. It decreases in the lesion as well, just not as much as the green fluorescence. We call this imaging modality (green fluorescence channel plus red fluorescence channel) the LIF mode. LIF stands for light-induced fluorescence. Subsequent technology development, thus, has created better reference images that are based on red near-IR band tissue reflectance signals, which vary much less between lesion and normal tissue (Zeng et al. 1998). We call this imaging modality (green fluorescence channel plus red or near-IR reflectance channel) the LIFR mode, with R standing for reflectance.

Figure 5.2 shows an endoscopic fluorescence imaging system for gastrointestinal (GI) cancer detection that is capable of both the LIF mode and LIFR mode imaging as well as white light (WL) imaging (Zeng et al. 1998). It consists of an illumination console, image acquisition hardware, a PC-based control console, and a color display monitor. All three types of illumination are provided by a filtered mercury arc lamp light source. The filtered light from the arc lamp was focused onto the illumination fiber bundle of the endoscope and transmitted into the GI tract to illuminate the tissue. Reflected light or autofluorescence light is collected by the imaging fiber bundle of the endoscope and transmitted to either the RGB camera or the fluorescence camera coupled to the eyepiece of the endoscope for image acquisition. The video output signals of the cameras are digitized by the control center and displayed in real time on the color monitor. The fluorescence image is displayed in pseudo color. Detailed descriptions of the system follow.

5.1.1 Light Source

A mercury arc lamp light source with a filter wheel provides three types of illumination for the three imaging modes. For *WL imaging mode*, a customized mercury line

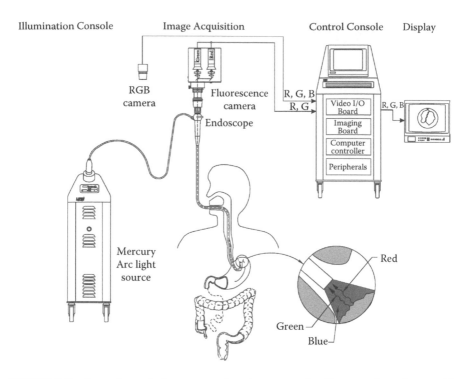

FIGURE 5.2   An endoscopic fluorescence imaging system for GI cancer detection capable of three imaging modalities: WL imaging mode, LIF imaging mode, and LIFR imaging mode. (Reproduced with permission from Zeng, H. et al. 1998. *Bioimaging* 6: 151–165.)

attenuation filter was used to obtain spectrally uniform white light illumination. For the *LIF imaging mode*, pure blue light is needed for fluorescence excitation. A 450 nm EFSP filter (Omega Optical Inc.) is used to filter out light above 450 nm, while UV light has been filtered out by a 399 nm LP filter. High out-of-band rejection is critical because light leakage at wavelengths longer than 450 nm would interfere with the autofluorescence signal. The blue light obtained is within the wavelength band from 400 nm to 450 nm. Blue light power of 50 mW to 80 mW was obtained out of the GI-endoscopes. For the *LIFR imaging mode*, the tissue has to be illuminated by both blue light and red near-IR light, simultaneously. The blue light excites fluorescence for the green channel fluorescence imaging, while the red near-IR light is reflected from the tissue and are used for the red channel reflectance imaging. The same 450 nm EFSP filter as used for LIF imaging mode was modified to generate the same blue light and also pass light of wavelengths longer than 590 nm. The filter then allows blue light in the 400 nm to 450 nm band and red near-IR light above 590 nm to be generated by the light source.

5.1.2 Image Acquisition

For the white light imaging mode, a SONY CCD RGB video camera (XC-77R) was used to capture the white light images. The camera outputs standard R, G, B video

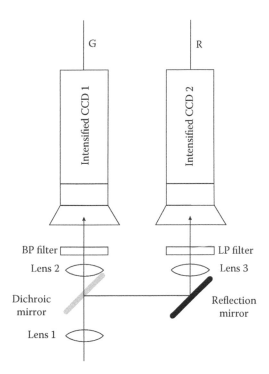

FIGURE 5.3 Configuration of the fluorescence camera used for both LIF imaging mode and LIFR imaging mode. A dichroic mirror and filters are used to separate the light into green and red wavelength bands. Two intensified CCD cameras are used to capture images in the two wavelength bands. (Reproduced with permission from Zeng, H. et al. 1998. *Bioimaging* 6: 151–165.)

signals. For fluorescence imaging, both LIF and LIFR modes, the same image acquisition hardware is used. Figure 5.3 shows the configuration of the camera. The light beam out of the endoscope eyepiece is collimated by the adapter lens 1. A dichroic mirror then separates the collimated light into two beams: one beam with wavelengths shorter than 600 nm transmitted through the mirror, and the other beam with wavelengths longer than 600 nm, reflected by the mirror and turned 90° toward the right side. The transmitted beam was focused by lens 2 onto the intensified CCD 1 to form the green channel image, generating the G video signal. The BP filter only passes wavelengths between 490 nm to 560 nm to the intensified CCD 1. The reflected beam is reflected by a reflection mirror again and turned 90° back to the original direction, where it is then focused by lens 3 onto the intensified CCD 2 to form the red channel image, generating the R signal. The LP filter only passes wavelengths above 630 nm. The two CCD cameras are genlocked to produce synchronized R and G video signals.

The video signals are feed into a video I/O board as shown in Figure 5.2, which handles switching of the video signals. In the white light imaging mode, the video I/O board then sends the R, G, B signals to the corresponding R, G, B input channels of the imaging board for digitization. In the fluorescence (LIF and LIFR) modes, the

R signal was sent to the R input channel of the imaging board, while the G signal was sent to both the G and B input channels of the imaging board for digitization. The computer and the imaging board then process the images in real time and output the processed R, G, B signals to a monitor for image display. The R, G, B images are directly overlaid on the monitor and presented to the physician in real time. The physician perceives a color image in which the color of each pixel is determined by the red/green-blue ratios calculated by the physician's vision system instead of the computer. The direct overlay of the R, G, B images also allow the physician to see the morphological structures and associated contextual information of the area being examined. A dedicated software system controls the operation of the whole system. When switching imaging modalities, the computer signals the filter wheel of the light source to achieve the desired illumination wavelengths, as well as sets the desired gain relationship and initial gain settings of the two intensified CCD cameras according to the imaging modes: LIF or LIFR.

5.1.3 PRINCIPLES OF FLUORESCENCE IMAGE FORMATION

For *LIF imaging mode*, the system produces a real-time pseudo color display based on the red/green images of the autofluorescence signal. The R signal is mapped to the R channel of the monitor, and the G signal is mapped to the G and B channels of the monitor. The images clearly delineate the suspicious areas (with a higher red/green-blue display ratio) from the normal tissue (with a lower red/green-blue display ratio) for biopsy. The colors of the pseudo color image are mainly determined by the red/green ratios of the autofluorescence signal. A dysplastic or neoplastic lesion appears reddish because its autofluorescence emission has a higher red/green ratio than does the normal tissue, while the normal tissue appears cyan.

Figure 5.4 shows the LIF image (A) and the WL image (B) of a polyp in the sigmoid colon. From the white light image, one could not tell if the polyp is benign or malignant. The LIF image showed a reddish patch on the polyp, while the surrounding normal tissue is in cyan color. Biopsy results of the reddish area on the polyp proved that this is a tubular adenoma with moderate dysplasia. Figure 5.4 also shows the distribution of fluorescence intensities (C) and the ratio (D) of red fluorescence intensity to green fluorescence intensity (red/green ratio) along a line across the LIF image in (A). The line starts at the "x" mark on the up left corner and ends at the "x" mark on the bottom right corner. The fluorescence intensity at a specific pixel depends on both the intrinsic fluorescence quantum yield of the tissue at the pixel and the distance from the tissue at that pixel to the endoscope. Both the green and red intensities start with high values in Figure 5.4 because the tissue surface is elevated in the corresponding pixels. They then gradually decrease until the edge of the polyp. At the elevated polyp, the tissue is closer to the endoscope, therefore, both the green signal and the red signal tend to increase. However, the intrinsic fluorescence quantum yield decreases due to the presentation of dysplasia, and the red fluorescence quantum yield decreases much less than does the green. The combination of these two effects results in the green signal being increased slightly at the polyp, while the red signal is increased sharply and significantly. At the end of the polyp, the tissue turns normal again, and we see an increase in the green signal accompanied by

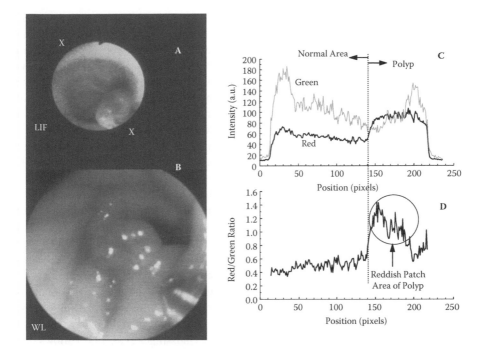

FIGURE 5.4 (See color insert) LIF image (A) and white light image (B) of a tubular adenoma polyp in the sigmoid colon. Part of the polyp appears reddish in the LIF image. Profiles of fluorescence intensities (C) and Red/Green ratio (D) along a line across the LIF image (A). The line starts at the "x" mark on the up left corner and ends at the "x" mark on the bottom right corner. (Reproduced with permission from Zeng, H. et al. 1998. *Bioimaging* 6: 151–165.)

a slight decrease in the red signal. Therefore, using either the single channel green fluorescence or the red fluorescence, one cannot detect a cancerous or precancerous lesion reliably due to dependence of the detected fluorescence signal on the endoscopic measurement geometry. Figure 5.4D shows the red/green ratio distribution at the same pixel positions as in Figure 5.4C. This ratio signal is almost constant before reaching the polyp. When entering the polyp, the ratio signal increased two folds due to the presentation of dysplasia. At the end of the polyp the ratio decreases again denoting normal tissues. Here, the red/green ratio cancels the geometry dependence of the fluorescence signal detected because the red signal and green signal have the same geometric dependence.

In the *LIFR imaging mode*, the differences in green fluorescence intensity between normal tissue and dysplastic or neoplastic lesions were exploited for diagnosis. However, if only the green fluorescence signal were used to form a single channel gray image (which is the case for the SAFE 1000 system from Pentax), the fluorescence intensity at each pixel and the intensity 2-D distribution would be affected by the nonuniform distribution of the illumination intensity on the tissue surface and the nonuniform fluorescence collection efficiency for each pixel caused by curved surfaces, folds, polyps, the angle of the endoscope relative to the tissue surface, and distance effects. Therefore, the image would not represent the actual fluorescence distribution on the tissue surface. A reference image is needed to correct for the nonuniform illumination distribution and nonuniform collection efficiency. By using a reference image for normalization, the user can distinguish between changes in intensity and changes in color of the image. The red fluorescence image serves as such a reference in the LIF imaging mode. However, because the red fluorescence intensity often changes from normal to abnormal tissue, the normalization is not ideal. It turns out that the change in the reflectance in the red near-infrared between normal and abnormal tissue is typically much less than the change in red autofluorescence. (Otherwise, the conventional white light imaging would contain enough information to differentiate the lesions from normal tissue.) Therefore, in the LIFR imaging mode, the red reflectance image serves as the reference to correct for the nonuniform illumination distribution and nonuniform collection efficiency over the imaged tissue area. The red reflectance image makes an improved reference because the change in red near-infrared reflectance from normal to abnormal tissue is much smaller than the changes in red autofluorescence. The red illumination light and blue illumination light have very similar spatial distributions. The collection efficiencies for scattered red light and for reemitted fluorescence green light are also similar. The pseudo image obtained accurately represents the intensity distribution of the auto-fluorescence emission from a lesion and its surrounding normal tissue. Additionally, using red near-infrared reflectance images here instead of green or blue reflectance images (such as in the Storz, D-Light/AF system, Häussinger et al. 1999) allows one to avoid artifacts due to blood absorption.

Figure 5.5 shows all the three types of images (LIF, LIFR, WL) from a dysplastic lesion of the esophagus. The lesion appears reddish on both the LIF image and the LIFR image. However, the LIFR image has a better signal-to-noise ratio and gives better contrast between the diseased tissue and the normal tissue. This improvement is partially due to the slightly higher blue excitation light intensity in the LIFR mode than in the LIF mode. The much greater signal-to-noise ratio of the red reflectance image in the LIFR mode compared to the poorer signal-to-noise ratio of the red fluorescence signal in the LIF mode is the dominant factor leading to the better signal-to-noise ratio in the LIFR mode. Figure 5.5 (D) and (E) show the separated green channel image and red channel image of the LIFR image in (B). The green channel image shows large variations in intensity, while the red channel image has quite uniform intensity distribution. If one only looks at the single channel fluorescence image (the green channel image), he/she cannot distinguish an empty hole or shaded area from an area with very low intensity of fluorescence signal. They are both dark spots on the image. However, on the red reflectance image, the area with low fluorescence intensity can be as bright as its surrounding area, while the empty hole or shaded area remains dark. Therefore, the two channel composite image as seen in Figure 5.5 (B) shows the reduced fluorescence signal area as reddish, where the empty hole or shaded area looks dark. This is an added advantage of the LIFR imaging modality compared to the single-channel fluorescence imaging modality (such as in the Pentax SAFE 1000 system). Another advantage of the LIFR imaging modality compared to the LIF modality is that the intensified CCD 2 in Figure 5.3 can be replaced with an ordinary CCD camera, reducing the cost of the system.

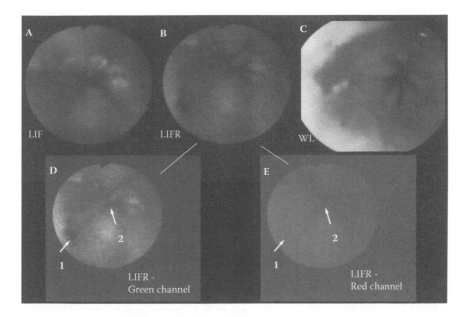

FIGURE 5.5 (See color insert) LIF image (A), LIFR image (B), and WL image (C) of an esophagus dysplasia. The separate green channel image (D) and red channel image (E) of the LIFR image are also shown. (Reproduced with permission from Zeng, H. et al. 1998. *Bioimaging* 6: 151–165.)

5.2 OTHER IMPROVEMENTS

The LIFR modality is currently used in most commercial fluorescence endoscopy systems (Xillix Onco-LIFE, Wolf DAFE, Olympus AFI). Sensitive CCD cameras have replaced ICCDs (Storz D-Light/AF, Zeng et al. 2004a). Originally implemented on fiber optic endoscopes, fluorescence endoscopy has also been implemented on video endoscopes with CCD at the tip (Pentax SAFE-3000 system, Olympus AFI system, Chiyo et al. 2005; Ikeda et al. 2006). The Pentax SAFE 3000 device realized dual image display of side-by-side white light and autoflourescence images (Ikeda et al. 2006; Lee et al. 2007).

Another important advancement is an integrated endoscopy system for simultaneous imaging and spectroscopy (Zeng et al. 2004a). The system employs the same CCD camera for both WL and LIFR image acquisition. It also realized spectral measurements of an interested tissue area through the camera without introducing a fiber optic catheter through the biopsy channel of the endoscope. This system utilizes simultaneous spectral measurements (reflectance and fluorescence) to reduce the false positives of fluorescence imaging. Details will be discussed in Chapter 6.

5.3 ORIGINS OF THE SIGNALS

An autofluorescence signal provides information about the biochemical composition and metabolic state of biological tissues. Most native tissue fluorophores are associated with the tissue matrix or are involved in cellular metabolic processes. Collagen and elastin are the dominant structural fluorophores and their fluorescence properties are very sensitive to cross-links between amino acids. Fluorophores involved in cellular metabolism include nicotinamide adenine dinucleotide (NADH) and flavins. Other fluorophores include tyrosine, tryptophan, various porphyrins, and lipopigments. Tissue fluorescence properties is determined by the concentration of these fluorophores, the distinctive excitation and emission spectra of each fluorophore, spatial distribution of various fluorophores in the tissue, and the metabolic state of the fluorophores. The detected fluorescence signals from the tissue surface have also been modified by (1) the reabsorption of the fluorescence photons by various chromophores in the tissue, such as hemoglobin and melanin and (2) the scattering of the fluorescence photons by various tissue microstructures related to tissue architecture and morphology. The *in vivo* tissue fluorescence measurement process has been modeled carefully and in detail using Monte Carlo simulation (for examples, see Zeng et al. 1997).

Using fluorescence for lung cancer detection as an example, when the bronchial surface is illuminated by blue light, normal tissues fluoresce strongly in the green. As the bronchial epithelium changes from normal to dysplasia, and then to carcinoma *in situ* and invasive cancer, there is a progressive decrease in green autofluorescence but proportionately less decrease in red fluorescence intensity (Hung et al. 1991). This may be attributed to several factors related to tissue morphologic and metabolic changes.

1. With blue light excitation, the fluorescence quantum yield in the submucosa is an order of magnitude higher than the epithelium (Qu et al. 1995). There is a decrease in extracellular matrix such as collagen and elastin in dysplasia and cancer, thus leading to decreasing green and red fluorescence. On the other hand, fluorescence emission from surrounding normal tissue diffuses into the lesion site with longer wavelength red light encounter less attenuation due to decreasing tissue absorption and scattering coefficients with increasing wavelengths.
2. The increase in the number of cells associated with dysplasia or cancer decreases the fluorescence detected in the bronchial surface due to absorption and scattering of the excitation light and re-absorption and scattering of the fluorescence light by the thickened epithelium.
3. The microvascular density is increased in dysplastic and cancer tissues. The presence of an increased concentration and distribution of hemoglobin results in increased absorption of the blue excitation light and fluorescence light.
4. There is a reduction in the amount of flavins and NADH in premalignant and malignant cells. Other factors such as pH and oxygenation may also alter the fluorescence quantum yield, spectral peak positions and line widths (Wolfbeis 1973).

5.4   ADVANTAGES AND LIMITATIONS

Principally, an increase in signal over a zero background signal is measured in fluorescence imaging, while in white light (reflectance) imaging, the analogous quantity, absorbed or scattered light, is measured indirectly as the difference between the incident and the reflected light. This small decrease in the intensity of a very large signal is measured in reflectance techniques leading to a correspondingly large loss in sensitivity (Guilbaut 1973; Zeng 1993). This physical advantage, coupled with the high sensitivity of fluorescence to the biochemical environment of the fluorophores makes fluorescence imaging much more sensitive for cancer detection than white light imaging does. A multicenter clinical trial for lung cancer detection has demonstrated that the relative sensitivity of fluorescence endoscopy versus white light endoscopy was 6.3 for detection of intraepithelial neoplastic lesions and 2.71 when invasive carcinomas were also included (Lam et al. 1998).

Although fluorescence endoscopy is the field imaging modality with the highest diagnostic sensitivity for early cancer detection, its diagnostic specificity is decreased (high false positives). For example, sometimes inflammatory tissues are hard to be differentiated from cancerous tissues. Therefore, it is necessary to use point spectroscopy and "point" microscopy methods to help decide if a suspicious lesion localized by fluorescence endoscopy should be biopsied or not. This will significantly reduce the number of false positive biopsies. These point measurement methods will be discussed in detail in Chapters 6–10.

REFERENCES

Anderson, R.R. and Parrish, J.A. 1982. Optical properties of human skin. In *The Science of Photomedicine,* ed. Regan, J.D. and Parrish, J.A. New York and London: Plenum Press, pp. 147–194.

Anderson, P.S., Montan, S., and Svanberg, S. 1987. Multispectral system for medical fluorescence imaging. *IEEE J. Quantum Electron.* 33: 1798–1805.

Baumgartner, R., Fisslinger, H., Jocham, D., Lenz, H., Ruprecht, L., Stepp, H., and Unsold, E. 1987. A fluorescence imaging device for endoscopic detection of early stage cancer— Instrumental and experimental studies. *Photochem. Photobiol.* 46: 759–763.

Caplan, R.M. 1967. Medical uses of the Wood's lamp. *JAMA*, 202: 1035–1038.

Chiyo, M, Shibuya, K, Hoshino, H, Yasufuku, K., Sekine, Y., Iizasa, T., Hiroshima, K., and Fujisawa, T. 2005. Effective detection of bronchial preinvasive lesions by a new autofluorescence imaging bronchovideoscope system. *Lung Cancer* 48: 307–313.

Goujon, D., Zellweger, M., Radu, A, Weber, B., van den Bergh, H., Monnier, P., and Wagnieres, G. 2003. *In vivo* autofluorescence imaging of early lung cancers in the human tracheobronchial tree with a spectrally optimized system. *J. Biomed. Optics* 8: 17–25.

Guilbault, G.G. 1973. *Practical Fluorescence—Theory, Methods, and Techniques.* New York: Marcel Dekker.

Häussinger, K., Stanzel, F., Huber, R.M., Pichler, J., and Stepp, H. 1999. Autofluorescence detection of bronchial tumors with the D-Light/AF. *Diagn. Ther. Endosc.* 5: 105–112.

Hirano, T., Ishizuka, M., Suzuki, K., Ishida, K., Suzuki, S., Miyaki, S., Honma, A., Suzuki, M., Aizwa, K., Kato, H., and Hayata, Y. 1989. Photodynamic cancer diagnosis and treatment system consisting of pulse lasers and an endoscopic spectro-image analysis. *Lasers Life Sci.* 3: 1–18.

Hung, J., Lam, S., LeRiche, J., and Palcic, B. 1991. Autofluorescence of normal and malignant bronchial tissue. *Laser Surg. Med.* 11: 99–105.

Ikeda, N., Honda, H., Hayashi, A., Usuda, J., Kato, Y., Tsuboi, M., Ohira, T., Hirano, T., Kato, H., Serizawa, H., and Aoki, Y. 2006. Early detection of bronchial lesions using newly developed video endoscopy-based autofluorescence bronchoscopy. *Lung Cancer* 52: 21–27.

Lam, S., Palcic, B., McLean, D., Hung, J., Pon, A., and Profio, A.E. 1990. Detection of low dose Photofrin II. *Chest* 97: 333–337.

Lam, S., MacAulay, C., Hung, J., leRiche, J., Profio, A., and Palcic, B. 1993. Detection of dysplasia and carcinoma *in situ* with a lung imaging fluorescence endoscope device. *J. Thorac. Cardiovasc. Surg.* 105: 1035–1040.

Lam, S., Kennedy, T., Unger, M., Miller, Y.E., Gelmont, D., Rusch, V., Gipe, B., Howard, D., LeRiche, J.C., Coldman, A., and Gazdar, A.F. 1998. Localization of bronchial intraepithelial neoplastic lesions by fluorescence bronchoscopy, *Chest* 113: 696–702.

Lee P., Brokx, H.A.P., Postmus, P.E., and Sutedja, T. 2007. Dual digital video-autofluorescence imaging for detection of preneoplastic lesions. *Lung Cancer* 58: 44–49.

Montan, S., Svanberg, K., and Svanberg, S. 1985. Multicolor imaging and contrast enhancement in cancer–tumor localization using laser-induced fluorescence in Hematoporphyrin-derivative-bearing tissue. *Opt. Lett.* 10: 56–58.

Palcic, B., Lam, S., Hung, J., and MacAulay, C. 1991. Detection and localization of early lung cancer by imaging techniques. *Chest* 99: 742–743.

Potter, W.R. and Mang, T.S. 1984. Photofrin II levels by *in vivo* fluorescence photometry. In *Porphyrin Localization and Treatment of Tumours*, eds. Doiron, D.R. and Gomer, C.J. New York: Alan R. Liss, pp. 177–186.

Profio, A.E., Doiron, D.R., and King, E.G. 1979. Laser fluorescence bronchoscope for localization of occult lung tumors. *Med. Phys.* 6: 523–525.

Qu, J., MacAulay, C., Lam, S., and Palcic, B. 1995. Laser-induced fluorescence spectroscopy at endoscopy: Tissue optics, Monte Carlo modeling, and *in vivo* measurements. *Opt. Eng.* 34: 3334–3343.

Wagnieres, G., Depeursinge, C.H., Monnier, P.H., Savary, M, Cornaz, P., Chatelain, A., and van den Bergh, H. 1990. Photodetection of early cancer by laser induced fluorescence of a tumour-selective dye: Apparatus design and realization. *SPIE Proc.* 1203: 43–52.

Wolfbeis, O. 1973. Fluorescence of organic natural products. In: *Molecular Luminescence Spectroscopy*, ed. Schulman, S.G., Vol. 1. New York: John Wiley & Sons, pp. 167–370.

Zeng, H. 1993. Human Skin Optical Properties and Autofluorescence Decay Dynamics. PhD Thesis, University of British Columbia.

Zeng, H., MacAulay, C., McLean, D.I., and Palcic, B. 1997. Reconstruction of *in vivo* skin autofluorescence spectrum from microscopic properties by Monte Carlo simulation. *J. Photochem. Photobiol. B: Biol.* 38: 234–240.

Zeng, H., Weiss, A., Cline, R., and MacAulay, C. 1998. Real time endoscopic fluorescence imaging for early cancer detection in the gastrointestinal tract. *Bioimaging* 6: 151–165.

Zeng, H., Petek, M., Zorman, M.T., McWilliams, A., Palcic, B., and Lam, S. 2004a. Integrated endoscopy system for simultaneous imaging and spectroscopy for early lung cancer detection. *Opt. Lett.* 29: 587–589.

Zeng, H., McWilliams, A., and Lam, S. 2004b. Optical spectroscopy and imaging for early lung cancer detection: A review. *Photodiagn. Photodyn. Ther.* 1: 111–122.

Section III

Endoscopic Point Spectroscopy

6 Endoscopic Reflectance and Fluorescence Spectroscopy

Haishan Zeng
British Columbia Cancer Agency Research Centre,
Vancouver, British Columbia, Canada

CONTENTS

Different from endoscopy field imaging modalities, which measure light intensity distributions of very few wavebands over a centimeters size field of view of the tissue under examination, endoscopic spectroscopy measures light intensity as a function of wavelength or frequency over a tissue area (spot/point) of millimeter size or smaller. For diagnostic purposes, field imaging and point spectroscopy are complimentary. Imaging facilitates quick survey and localization of suspicious areas/spots over a large field. Spectroscopy can be used to perform detailed analysis of suspicious area/spots identified by imaging. Imaging determines diagnostic sensitivity, while spectroscopy can improve diagnostic specificity, reducing false positive biopsies. Sometimes, spectroscopy measurements over many randomly selected spots can also improve diagnostic sensitivity, such as in the case of Barrett's esophagus management. Combination of imaging and spectroscopy often leads to improvement of the overall diagnostic performance. Not all optical spectroscopy modalities can be conveniently implemented for endoscopic applications due to the technical challenges associated with endoscopy. We will focus on the three most common spectroscopy modalities: reflectance spectroscopy (including light scattering spectroscopy) and fluorescence spectroscopy in this chapter, and Raman spectroscopy in the next chapter.

6.1   BASICS OF REFLECTANCE AND FLUORESCENCE SPECTROSCOPY THROUGH ENDOSCOPY

The most common way of measuring reflectance and fluorescence spectra through an endoscope is utilizing a six-one bifurcated fiber bundle catheter that can pass through the endoscope instrument channel to be placed close to the tissue or in contact with the tissue surface. Figure 6.1 shows the schematics of such a system. The common distal end of the fiber bundle has a center fiber and six surrounding fibers that are separated into two branches at the proximal end of the catheter. The center fiber can be used for illumination, while the surrounding fibers are for reflectance or fluorescence light collection, or vice versa. Other varieties of fiber catheters can be found in a review paper by Utzinger and Richards-Kortum (2003).

The light source in the system provides illumination light focused into the illumination fiber/fibers for the desired spectral measurements. A broad-band white light source is used for reflectance measurements in the visible and/or near IR wavelength range. Example light sources include QTH (quartz tungsten halogen) lamp, high-pressure arc lamp (Xe, mercury, etc.), metal halide lamp. A long pass filter may be used to block the UV components from reaching the tissue that are carcinogenic. For fluorescence excitation, a laser (often with blue wavelengths) can be used. A laser line filter may be used to purify the excitation light if the laser has noises in the fluorescence wavelength range. Example lasers used for tissue fluorescence spectral measurements include Kr+ laser (405 nm), He-Cd laser (442 nm), Ar+ laser (488 nm), and diode laser (405 nm, 440 nm etc.). Alternatively, high pressure arc lamp or metal halide lamp can also be used for fluorescence excitation by coupling with a monochromator and/or high performance band pass filter. In this case, the same light

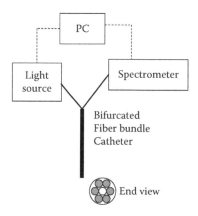

FIGURE 6.1   Schematics of reflectance and fluorescence spectral measurement setup for endoscopy.

source can be used for both fluorescence excitation and reflectance illumination by switching between filters (Zeng et al. 1993).

The spectrometer is interfaced with the collection fiber/fibers branch of the catheter to perform spectral analysis. Since rapid data acquisition (<1 second) is critical for clinical endoscopy applications, a diffraction grating–CCD array detector-based dispersive type spectrometer is the way of choice. Since the tissue reflectance and fluorescence signals are relatively strong and the spectral shapes are relatively smooth, no sharp peaks, low end spectrometers with resolutions of a few nanometers are good enough. Even a $2,000 palm-size miniature spectrometer (e.g., Model USB2000, Ocean Optics) is sufficient for many applications. A single collection fiber catheter is often used with miniature spectrometers since it only takes single fiber as input. If multiple fibers are used for signal collection, they are arranged in a linear pattern in the fiber connector (such as a SMA connector) to be compatible with the linear entrance slit of the full-size spectrometer. For reflectance spectral measurements, the collection fiber/fibers can be directly connected to the spectrometer without further filtering. For fluorescence measurements, a long pass filter is required to block the reflected excitation light from entering the spectrometer. For a full-size spectrometer, a filter holder with a collimating lens and a focusing lens can be mounted in front of the entrance slit of the spectrograph to accommodate the long pass filter (Hung et al. 1991). For a miniature spectrometer, an inline filter holder can be used, which also consists of a collimating lens and a focus lens to relay the light from the collection fiber to another fiber connected to the spectrometer (Zeng et al. 1995). The long pass filter is placed in between the two lenses. By removing and inserting the filter, both reflectance and fluorescence spectral measurements can be performed with the same spectrometer (Zeng et al. 1993, 1995).

A PC computer is used to control/synchronize the light source and the spectrometer, store, and process the acquired spectral data. Unless pulsed light source and gating electronics are used, the PC also controls the imaging light source of the endoscope system to make sure that it is momentarily off during the spectral acquisition.

Spectral wavelength calibration is performed using known atomic spectral lines of a low power, low pressure mercury-argon lamp. For fluorescence measurements, intensity calibration should be made using a standard lamp (e.g., RS-10, EG&G Gamma Scientific) and the supplied spectral data. The intensity calibration enables the correction of the wavelength dependence of the response of the whole system.

Reflectance spectrum is obtained by taking two readings: one from the tissue and one from a reflectance reference standard, which has near 100% reflectivity over the measurement wavelength range. Taking a ratio of the tissue spectrum divided by the reference spectrum generates the reflectance spectrum. Reflectance standards are available from a couple of suppliers (e.g., SRS-99, Labsphere, Inc.).

6.2 SPECTRAL MEASUREMENT GEOMETRIES FOR OPTIMIZING IMAGING SYSTEM DESIGN

Although the bifurcated fiber bundle catheter is a popular setup in endoscopic spectral measurements, the illumination light distribution in tissue and the interrogation tissue volume are very different from the imaging measurement geometry. This is

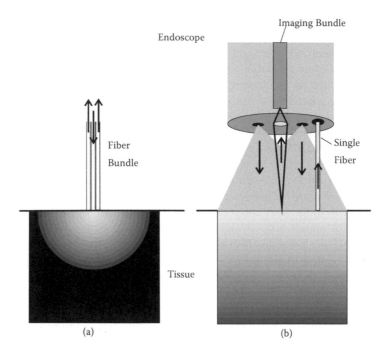

FIGURE 6.2 Comparison of two different spectral measurement geometries and corresponding illumination light distribution inside the tissue. (A) Bifurcated fiber bundle: central fiber for illumination, surrounding fibers for collection. (B) Imaging-like measurement geometry: endoscope light guides for illumination, single fiber through the instrument channel for collection.

illustrated in Figure 6.2. The fiber bundle measurement geometry is more like a point illumination; the light fluence has a hot spot near the tissue surface and close to the fiber. The fluence decreases radially surrounding the hot spot. In the situation of imaging, we have a broad beam illumination pattern; the light fluence distribution has a flat profile parallel to the tissue surface. The fluence decreases with increasing depth inside the tissue. On average, light penetration depth/spectral measurement interrogation tissue volume in the imaging geometry (Figure 6.2B) are deeper than the bifurcated fiber bundle spectral measurement geometry (Figure 6.2A). Therefore, if we want to use the spectral data to optimize the imaging system—that is, find out the best wavelength bands for fluorescence imaging—we should measure the spectra using geometry similar to the imaging situation. That is broad beam illumination and point detection as illustrated in Figure 6.2B. Light from a point on the tissue surface are collected by the endoscope objective and focused to a single fiber in the imaging bundle. This illumination/detection geometry for spectral measurement can be realized by (1) employing the endoscope light guide to conduct the spectral measurement illumination light, which is implemented by using the same imaging light for spectral measurement, and (2) collecting the spectral signal from a point on the tissue surface through a single optical fiber passing through the endoscope instrument channel (Figure 6.2B). Hung et al. (1991) performed an endoscopic fluorescence spectral study of normal, premalignant, and malignant bronchial tissues using such an imaging-like geometry. The study results had served as the basis for the design of a fluorescence endoscopic imaging device for lung cancer detection (LIFE-Lung™, Xillix Technology, Inc.).

6.3 INTEGRATED ENDOSCOPY SYSTEM FOR SIMULTANEOUS IMAGING AND SPECTROSCOPY WITHOUT USING A FIBER CATHETER

Figure 6.2B also suggests the possibility of collecting spectral signal from a spot on the tissue surface using the imaging fiber bundle simultaneously with imaging if we can somehow separate the spectral signal from the imaging signal (one way is illustrated in Figure 6.6). With this idea in mind, we have successfully developed an integrated endoscopy system for simultaneous imaging and spectroscopy without using a fiber catheter for lung cancer detection (Zeng et al. 2004). Figure 6.3 shows the block diagram of the system. It consists of a special light source, an endoscope, a special camera unit, a spectrometer (USB2000, Ocean Optics), and a PC. The light source uses a Xenon arc lamp and provides both white light (400–700 nm) for white light (WL) imaging and reflectance spectral measurements and strong blue light (400–460 nm) plus weak near-IR (NIR) light (720–800 nm) for fluorescence (FL) imaging and fluorescence spectral measurements. The switching of these two illumination modes is achieved by a stepper motor that moves two different filters into the optical path. The illumination light guide of the endoscope is interfaced to the light source to illuminate the bronchial tree and the imaging bundle of the endoscope collects and relays the reflected and fluorescence light from the tissue surface to the camera unit (the camera module and the spectral attachment) for imaging and spectroscopy. An optical fiber carries the optical signal from a spot at the center of the image, which corresponds to an area of about 1 mm diameter at the tissue

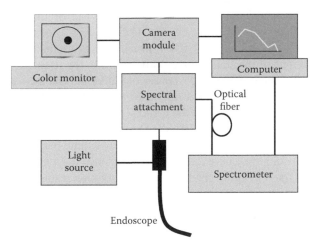

FIGURE 6.3 Block diagram of the integrated endoscopy system for simultaneous imaging and spectroscopy without using a fiber catheter.

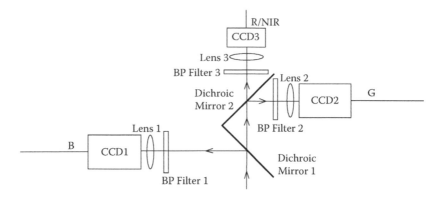

FIGURE 6.4   Optical layout of the camera module for both WL imaging and FL imaging. (Reproduced with permission from Zeng, H. et al. 2004. *Opt. Lett.* 29: 587–589.)

surface when the endoscope tip is about 10 mm away from the tissue surface, to the spectrometer for spectral analysis. The video image and spectrum (either in the WL mode or the FL mode) are simultaneously displayed at the computer monitor in live mode. A still image and spectrum at any interested time point during the endoscopy procedure can be captured and stored in the PC.

The optical layout of the camera module is shown in Figure 6.4. It consists of 3 imaging channels: blue channel (B) responsive to light in 400–500 nm, green channel (G) in 500–600 nm, and red/NIR channel (R/NIR) in 600–800 nm. Dichroic mirror 1 accepts the image beam entering the camera and reflects light below 500 nm, while transmitting light above 500 nm. Dichroic mirror 2 reflects light below 600 nm and transmits light above 600 nm. BP filter 1 passes light in 400–500 nm and lens 1 focuses the transmitted blue beam to form an image at CCD 1, which outputs standard B video signal. BP filter 2 passes light in 500–600 nm with an out-of-band rejection optical density (OD) larger than 5 and lens 2 focuses the transmitted green beam to form an image at CCD 2, which outputs a standard G video signal. BP filter 3 passes light in 600–800 nm with an out-of-band rejection OD>5 and lens 3 focuses the transmitted red/NIR beam to form an image at CCD 3, which outputs a standard R video signal.

The convolutions of the light source output spectra, the tissue responses to incident light, and the camera spectral responses of individual channels were utilized to generate the desired system spectral responses to facilitate both WL and FL imaging with the same set of sensors in the camera. Figure 6.5 outlines the convolution processes for both WL (a,b,c) and FL (d,e,f) imaging modes. Curve 1 in (a) is the light source output spectrum in WL mode, basically a quite flat profile in the visible wavelengths 400–700 nm, curve 2 the tissue reflectance response, curve 3 the spectral responses of the 3 imaging channels. The convolution results are depicted in (c). The B channel integrates reflected photons from 400 nm to 500 nm, the G channel from 500 to 600 nm, while the R channel is 600–700 nm.

In FL imaging mode (equivalent to the LIFR mode in Chapter 5), the light source output spectrum is shown in Figure 6.5d, curve 1, consisting of a strong blue band of 400–460 nm and a weak NIR band above 720 nm, curve 2 the tissue responses consisting of reflected blue light in 400–460 nm, reflected NIR above 720 nm, and the

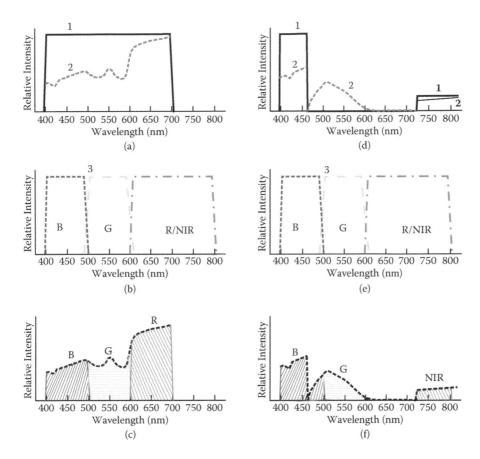

FIGURE 6.5 Convolution processes for realizing both WL imaging (a,b,c) and FL imaging (d,e,f) using the camera module of Figure 6.4. All curves 1, 2, and 3 are schematic representations only, not real data and not to scale. (Reproduced with permission from Zeng, H. et al. 2004. *Opt. Lett.* 29: 587–589.)

fluorescence emission in 460–700 nm. The convolution results are shown in (f), with the B channel integrating the reflected blue light in 400–460 nm and the fluorescence light in 460–500 nm, the G channel integrating fluorescence light in 500–600 nm, while the NIR channel integrating reflected NIR is 720–800 nm. The NIR channel reflectance image is used to normalize the G channel fluorescence image. The B channel is disabled for simpler visual interpretation of the FL image, which is the superimposition of the G channel fluorescence image and the NIR channel reflectance image. A cancerous lesion will appear reddish in the FL image.

This convolution approach of image formation has allowed us to perform both WL and FL images with a single 3-CCD camera without any mechanical switching, resulting in a more compact, less expansive, user friendlier, and lighter weight camera. The image mode switching is realized by changing between two filters in the light source. Of course, the camera is switched to different gain settings when the imaging mode is changed.

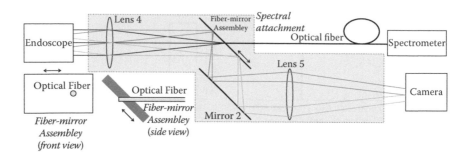

FIGURE 6.6 Optical layout of the spectral attachment. The front view and side view of the fiber-mirror assembly, a key component of the attachment, are also shown. (Reproduced with permission from Zeng, H. et al. 2004. *Opt. Lett.* 29: 587–589.)

Noncontact, catheter-free, simultaneous spectral measurement with imaging is realized by placing a specially designed spectral attachment between the endoscope eyepiece and the camera module. Figure 6.6 shows the optical layout of this attachment (shaded section). Light coming out of the endoscope is focused by lens 4 to form an interim image at the fiber-mirror assembly, a key component for spectral measurements. This assembly is fabricated by drilling a hole through a mirror, then mounting a 200-μm core-diameter taper fiber through the hole. A cover glass is attached to the mirror surface to prevent dust from accumulating on the image plan and affecting the image quality. The fiber is placed at an angle to the mirror surface so that its end face is perpendicular to the incident beam in order to collect all the light falling on the fiber. The fiber collects light from a 200-μm spot at the center of the image and transmits the photons to the spectrometer for spectral analysis. This taper fiber has a 100-μm core-diameter at the spectrometer end to achieve a spectral resolution of 5 nm. The remaining portions of the image surrounding the fiber are reflected to mirror 2, which in turn reflect the beam to lens 5. Lens 5 then collimates the light beam before reaching the camera for electronic image acquisition. The image will have a dark spot appearing at the center that originates from light being carried away by the fiber, showing where exactly the spectrum is acquired from on the image, serving perfectly as an automatic alignment mechanism. A perfect image without the fiber dark spot can be obtained by sliding the fiber-mirror assembly using a solenoid along the arrow direction shown in Figure 6.6 so that the fiber is moved outside the field of the interim image.

Figure 6.7 shows example images and spectral curves of a carcinoma *in situ* (CIS) lesion obtained with the system during a patient endoscopy procedure. Figure 6.7a is the WL image with the optical fiber almost outside of the interim image field; (b) is the WL image when the fiber is aligned with the CIS for spectral measurement and shows as a black spot; (c) shows the reflectance spectra obtained from the CIS and its surrounding normal tissue. The hemoglobin absorption at 540 nm and 580 nm are clearly discernable. The ratio spectrum of CIS divided by normal is about constant below 580 nm, but increases monotonically above 580 nm. (d) is the FL image with the fiber almost outside of the interim image field; (e) is the FL image when the fiber is aligned with the CIS for spectral measurement; (f) shows the fluorescence spectra from the CIS and its surrounding normal tissue. The spectrum has a maximum

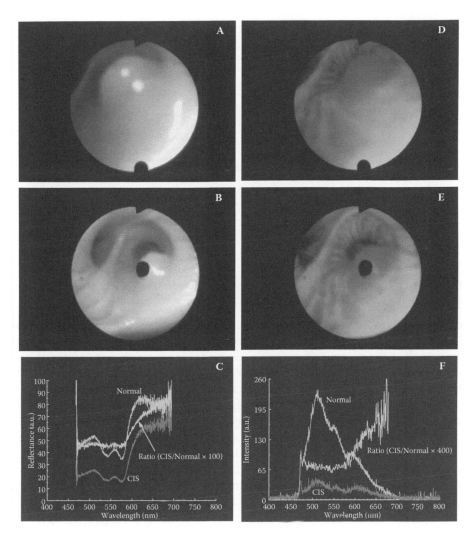

FIGURE 6.7 (See color insert) Example WL images, reflectance spectral curves (a,b,c) and FL images and spectral curves (d,e,f) of a *carcinoma in situ* lesion and its surrounding normal tissue. (Reproduced with permission from Zeng, H. et al. 2004. *Opt. Lett.* 29: 587–589.)

intensity around 512 nm and shows hemoglobin absorption at 540 nm and 580 nm. The fluorescence intensity of CIS is significantly lower than the normal tissue. The fluorescence ratio of CIS divided by normal is about constant below 580 nm, but increases monotonically above 580 nm, similar to the reflectance ratio spectrum.

6.4 TRANSPORT THEORY MODELING FOR SPECTRAL DATA INTERPRETATION FOR IMPROVING DIAGNOSTIC SPECIFICITY

The reflectance spectra and fluorescence spectra contains information about the tissue biochemistry/physiology and morphology. Figure 6.8 illustrates that cancer-related

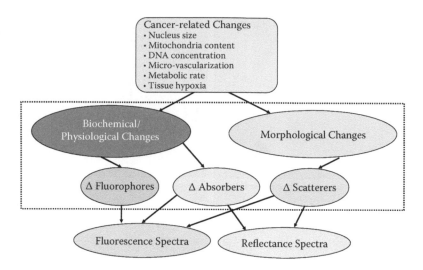

FIGURE 6.8 Illustration of cancer related changes alternate tissue biochemistry/physiology and morphology, and in turn, affect tissue fluorescence and reflectance spectra.

changes (e.g., nucleus size, mitochondria content, DNA concentration, microvascularization, metabolic rate, and tissue hypoxia) will alternate the tissue biochemistry and morphology, in turn, affect the tissue spectra *in vivo* such as measured by the system and method presented in Section 6.3 above. For improving lung cancer detection, we have developed transport theory modeling based methodology to relate the measured tissue spectral features with tissue physiology and morphology (Fawzy and Zeng 2006; Fawzy et al. 2006; Fawzy 2008). Related work has also been conducted by other groups, but for different spectral measurement geometry (Amelink et al. 2004; Bard et al. 2005a, 2005b). However, our modeling is simpler because our broad beam illumination generates a light distribution dependent of tissue depth only. This allowed us to create a one-dimensional (1-D) model. Our modeling on fluorescence spectra did not yield as good results as on reflectance spectra (Fawzy and Zeng 2008). The following discussion will be on reflectance spectra only.

6.4.1 Clinical Measurements

The *in vivo* reflectance signal measured from the tissue, $I_{m1}(\lambda)$, can be described as follows:

$$I_{m1}(\lambda) = a_1 I(\lambda) + b_1 I(\lambda) R_{tm}(\lambda) \tag{6.1}$$

where $I(\lambda)$ is the white light illumination reached the measurement spot on tissue surface, a_1 is a constant related to the efficiency by which the tissue-surface specular reflection was collected by the system, b_1 is a constant related to the efficiency of collecting diffuse reflected light from tissue by the measuring system and $R_{tm}(\lambda)$ is the true tissue diffuse reflectance to be derived. The reflectance signal measured from tissue $I_{m1}(\lambda)$ was divided by the reflectance signal measured from a reflectance standard disc to account for the illumination spectrum $I(\lambda)$. The reflectance signal measured from the standard disc, $I_{m2}(\lambda)$, can be described as:

$$I_{m2}(\lambda) = a_2 I(\lambda) + b_2 I(\lambda) R_s \tag{6.2}$$

where a_2 is a constant related to the efficiency by which the specular reflection was collected by the system. b_2 is a constant related to the efficiency by which the diffuse reflection signal was collected, and R_s is the reflectivity of the standard disc, which is a constant across the whole visible wavelength range. Dividing Equation (6.1) and Equation (6.2) and rearranging the equation, we obtained the following:

$$R_m(\lambda) = \frac{I_{m1}(\lambda)}{I_{m2}(\lambda)} = a_0 + b_0 R_{tm}(\lambda) \tag{6.3}$$

where, $R_m(\lambda)$ is the apparent reflectance spectra measured by our apparatus, $R_{tm}(\lambda)$ is the true tissue reflectance spectrum to be derived, and, a_0 and b_0 are additive offset and multiplicative factors respectively, which depend on the measurement conditions during each *in vivo* measurement performed. This includes the amount of specular reflection collected, the material of the standard disc, and the endoscope tip distance from the tissue during the measurement.

In the study (Fawzy et al. 2006), we conducted *in vivo* measurements of normal bronchial mucosa and both benign and malignant lesions on 22 patients. We obtained a total of 100 spectra. A biopsy sample was obtained for each measurement to classify the measured tissue site into normal, benign, or malignant. The pathology examination of biopsies confirmed that 21 reflectance spectra were from normal tissue sites, 29 from benign lesions (26 hyperplasia and 3 mild dysplasia), and 50 from malignant lesions (7 small cell lung cancer, 3 combined squamous cell carcinoma and non-small cell lung cancer, 30 non-small cell lung cancer, 10 adenocarcinoma). Our analysis was to develop algorithms to classify the spectra into two groups: (1) malignant lesions for tissue pathology conditions that were moderate dysplasia or worse and (2) normal tissue/benign lesions below moderate dysplasia. In clinical practice, group 1 lesions should be biopsied and treated (or monitored), while group 2 conditions could be left unattended. During routine clinical endoscopy examination all suspected malignant lesions (group 1) should be biopsied while group 2 conditions will not be biopsied. However, in this specially designed study, for each patient an extra biopsy (and corresponding spectral measurement) was taken randomly from either a normal-looking area or a suspected benign lesion so that we can assess the performance of the spectral diagnosis relatively independent of the performance of the imaging diagnosis. Five histopathology-confirmed malignant lesions were found by these random biopsies.

6.4.2 MODELING OF THE REFLECTANCE SPECTRA

We developed a theoretical model that links the tissue reflectance spectrum measured by our apparatus to specific tissue physiological and morphological parameters related to cancer changes. This was achieved by developing a light transport model with its optical absorption coefficient expressed in terms of the microvascular-related parameters and scattering coefficients expressed in terms of the tissue microstructure scatterer related parameters. The reflectance measurements performed by our apparatus with broad beam illumination and point collection can be represented by an equivalent 1-D measurement geometry (Fawzy et al. 2006). The light fluence distribution inside the tissue is a function of depth z only. Theoretically, the tissue reflectance spectra $R_t(\lambda)$ at each wavelength can be obtained using Fick's law:

$$R_t(\lambda) = \left.\frac{-j(z,\lambda)}{I_0}\right|_{z=0} = \gamma^{-1}\nabla\,\phi(z,\lambda)\big|_{z=0} \tag{6.4}$$

where ϕ is the light fluence spatial distribution, j is the diffuse flux, I_0 is the incident power, and γ is the diffusion constant which depends on the tissue's optical properties. The light fluence ϕ was obtained from the general diffusion approximation model (Venugopalan et al. 1998; Fawzy and Zeng 2006). For a CW plane source decaying exponentially in the z-direction, the general diffusion model is given by:

$$\nabla^2\phi(z) - \kappa_d^2\phi(z) = -\gamma S(z) \tag{6.5}$$

with

$$\kappa_d^2 = 3\mu_a\mu_{tr},\ \gamma = 3\mu_s^*(\mu_{tr} + g^*\mu_t^*)$$

where ϕ is the light fluence, $S(z)$ is the incident collimated source term, μ_{tr} is the transport attenuation coefficient equivalent to $[\mu_a + \mu_s(1 - g)]$, where μ_s and μ_a are the scattering and the absorption coefficients respectively. μ_t^* is the total attenuation coefficient and is equivalent to $[\mu_a + \mu_s^*]$. μ_s^* is the reduced scattering coefficient, which is equivalent to $\mu_s(1 - f)$, where f is the fraction of light scattered directly forward in the δ-Eddington approximation to the scattering phase function. g^* denotes the degree of asymmetry in the diffuse portion of the scattering. The values of f and g^* were related to the single scattering anisotropy: g from the matching of the second moment of the δ-Eddington phase function to the Henyey–Greenstein phase function, and are equivalent to g^2 and $g/(1 + g)$, respectively.

Since we were interested in reflectance spectral signals that were more affected by the superficial mucosa layer (~ 0.5 mm thickness), within which most early cancerous changes occur, we solved equation (6.5) for a two-layer tissue model (Fawzy et al. 2006) with the top layer thickness L setting to 0.5 mm. We solved equation (6.5) in the z-direction (1-D) for layer 1 and layer 2, using the refractive index mismatching boundary conditions at the air–tissue interface and the refractive index matching boundary conditions at the interface between the two tissue layers. By substituting the solution of Equation (6.5) into Equation (6.4) we obtained an expression for the tissue reflectance spectrum $R_t(\lambda)$ in terms of the absorption coefficient μ_a, the scattering coefficient μ_s, and the scattering anisotropy g.

The absorption coefficient, μ_a, was modeled in terms of the blood contents and the absorption coefficient of lung tissue measured *ex vivo* with blood drained out. Two parameters were used to describe the blood contents in tissue; the blood volume fraction, ρ, and the blood oxygen saturation, α. The absorption properties of lung tissue *in vivo* can be described by the following equations:

$$\mu_a(\lambda) = \mu_{blood}(\lambda)\rho + \mu_{ex\,vivo}(1-\rho),$$

$$\mu_{blood}(\lambda) = \alpha\mu_{HbO2} + (1-\alpha)\mu_{Hb}$$

(6.6)

where μ_{HbO2} and μ_{Hb} are the absorption coefficients for the oxy- and deoxyhemoglobin respectively. The *ex vivo* absorption coefficient, $\mu_{ex\,vivo}$, was obtained from the *ex vivo* lung tissue measurements made previously by Qu et al. 1994.

The scattering coefficient μ_s and the scattering anisotropy g were modeled in terms of the microstructure scatterer volume fractions and size distribution. The tissue scattering model was developed using the fractal approach, assuming that the tissue microstructures' refractive index variations can be approximated by a statistically equivalent volume of discrete microscattering particles with a constant refractive index but different sizes (Gelebart et al. 1996; Schmitt and Kumar 1998). The transport scattering coefficients for a bulk tissue can then be calculated by adding randomly the light waves scattered by each particle together (Fawzy and Zeng 2006; Schmitt and Kumar 1998). Thus, the transport scattering coefficient and the scattering anisotropy can be modeled using the following integral equations:

$$\mu_s(\lambda) = \int_0^\infty [Q(x,n,\lambda)]\frac{\eta(x)}{\upsilon(x)}dx$$

(6.7)

$$g(\lambda) = \frac{\int_0^\infty [g(x,n,\lambda)Q(x,n,\lambda)]\frac{\eta(x)}{\upsilon(x)}dx}{\int_0^\infty [Q(x,n,\lambda)]\frac{\eta(x)}{\upsilon(x)}dx}$$

(6.8)

where $Q(x)$ is the optical scattering cross section of individual particle with diameter x, refractive index (n), and wavelength (λ). $\upsilon(x)$ is the volume of the scattering particle with diameter x, $g(x)$ is the mean cosine of the scattering angles of single particle. For spherical microparticles, $Q(x)$ and $g(x)$ were calculated from Mie theory using the Mie-scattering code by Bohren and Huffman. The volume fraction distribution $\eta(x)$ were assumed to follow a Skewed logarithmic distribution (Fawzy and Zeng 2006; Schmitt and Kumar 1998):

$$\eta(x) = \delta C_0 x^{-\beta} \exp\left(-\frac{(\ln x - \ln x_m)^2}{2\sigma_m^2}\right)$$

(6.9)

where δ is the total volume fraction of all the scattering particles in tissue, β is the size-distribution parameter (fractal dimension), which determines the shape of the volume-fraction size distribution and is related directly to the size of the scattering particles, x_m and σ_m set the center and width of the distribution, respectively, and C_0 is a normalizing factor obtained from the condition

$$\delta = \int_0^\infty \eta(x)\,dx$$

The value of x_m was assumed equal to the geometrical mean of (0.05 μm) and (20 μm), which represent the limits of the scattering particles' range of diameters found typically in tissues (Fawzy and Zeng 2006). Thus, $x_m = [(0.05)(20.0)]^{1/2} = 1.0$. The width parameter σ_m was assumed to be a constant of 2.0 to match with the fractal scaling range of tissues (Schmitt and Kumar 1996). Having x_m and σ_m being set, the larger the value of β, the higher the contribution of the smaller size particles in the scattering particle size distribution function.

6.4.3 Inverse Algorithms

To obtain quantitative information about the blood volume fraction (ρ), the oxygen saturation parameter (α), the scattering volume fraction (δ), and the size-distribution parameter (β) from the measured *in vivo* reflectance spectra $R_m(\lambda)$, we developed a numerical inversion (fitting) algorithm based on a Newton-type iteration scheme through least-squares minimization of the function:

$$\chi^2 = \sum_i [R_m(\lambda_i) - (a_0 + b_0 R_t(\lambda_i))]^2 \qquad (6.10)$$

where $R_m(\lambda_i)$ is the reflectance measured at wavelength λ_i, $R_t(\lambda_i)$ is the computed reflectance at wavelength λ_i according to Equation (6.4). The iteration procedure were terminated when the χ^2 difference between two adjacent fittings became smaller than 0.01. The following parameters were used as free-fitting variables during the inversion process:

- *Blood volume fraction* (ρ) assumed to be the same for both tissue layers
- *Blood oxygen saturation parameter* (α) assumed to be the same for both tissue layers
- Scattering volume fraction in top and bottom layers (δ)
- *Size-distribution parameters* (β) in top and bottom layers
- Additive and multiplicative terms in Equation (6.3) (a_0, b_0).

Using the Marquardt-type regularization scheme (Press et al. 1992), we can obtain the updates of these parameters from the following system of equations:

$$(\zeta^T\zeta + v\mathbf{I})\Delta\tau = \zeta^T[R_m - (a_0 + b_0 R_t)] \tag{6.11}$$

where ζ is the Jacobian matrix, $\Delta\tau$ is the vector updates for the eight parameters (ρ, α, δ_1, δ_2, β_1, β_2, a_0, b_0), I is the identity matrix, and v could be a scalar or diagonal matrix. The Jacobian matrix ζ represents the sensitivity of the measured reflectance coefficients on the eight parameters and its elements were computed from the derivatives of $R_t(\lambda)$ with respect to these eight parameters. The inclusion of a_0 and b_0 in the fitting were essential to account for the specular reflection component and the diffuse reflectance measurement collection efficiency, which varied for each measurement and depended, among others, on the endoscope-tissue distance and angle. Thus, the true tissue reflectance (or called "corrected reflectance") $R_{tm}(\lambda)$ can then be extracted from the reflectance spectra measured by the apparatus, $R_m(\lambda)$, using the values of a_0 and b_0 that were obtained from the fitting procedure and substituting in Equation (6.3). Of course, all other parameters, ρ, α, δ_1, δ_2, β_1, and β_2 were also derived by this inverse algorithm.

6.4.4 STATISTICAL ANALYSIS

All the fitting results obtained from the 100 tissue spectral measurements were collected and saved in two groups (benign/normal and malignant) for statistical analysis. We used the STATISTICA software package (version 6, StatSoft Inc., Tulsa, OK) for the analysis. While we were not sure if the derived parameters follow normal distributions, the Kolmogorov–Smirnov two-sample test was chosen to evaluate the significance of differences between the two groups (normal tissue/benign lesions vs. malignant lesions) for each of the six parameters (ρ, α, δ_1, β_1, δ_2, β_2) obtained from our fitting results. Discriminant function analysis (DFA) was then applied to the identified diagnostically significant parameters to build diagnostic algorithms for tissue classification. DFA determined the discrimination function line that maximized the variance in the data between groups while minimizing the variance between members of the same group. The performance of the diagnostic algorithms rendered by the DFA models for correctly predicting the tissue status (i.e., normal/benign versus malignant) underlying each parameter set derived from the reflectance spectrum was estimated using the leave-one-out, cross-validation method on the whole data set (Dillion and Goldstein 1984; Lachenbruch and Mickey 1968).

6.4.5 RESULTS

The average of the corrected reflectance spectra (R_{tm}) for both the normal/benign group and the malignant group are shown in Figure 6.9. It shows that the average reflectance spectra of the normal/benign group have higher intensities in the measured wavelength range (470–700 nm) than the malignant group. These intensity differences are significantly larger for wavelengths above 600 nm. In addition the two hemoglobin absorption valleys around 540 nm and 580 nm are larger and more obvious on the normal/benign group spectral curve than on the malignant group spectral curve, consistent with tissue hypoxia in malignant lesions (Höckel and Vaupel 2001).

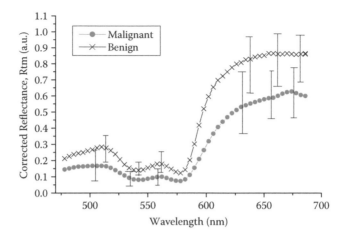

FIGURE 6.9 Average true reflectance spectrum of the 50 normal tissue/benign lesions versus that of the 50 malignant lesions. The error bars are show for a few data points to given an idea about the degree of the reflectance intensity overlap between the two groups. (Reproduced with permission from Fawzy, Y.S. et al. 2006. *J. Biomed. Opt.* 11: 044003[1–11].)

The average fitting parameters (ρ, α, δ_1, β_1, δ_2, β_2) and their standard deviations for the two groups are shown in Table 6.1. The mean value of the blood volume fraction was higher for malignant lesions (0.065 ± 0.03) compared to the benign lesions (0.032 ± 0.02). The mean value of the oxygen saturation parameter was reduced from 0.9 for benign lesions to 0.78 for malignant lesions. For the scattering parameters the mucosal layer showed moderate to significant changes between normal/benign tissues and malignant lesions with the mean values of δ_1 and β_1 for the benign lesions to be 0.077 and 0.97, respectively, compared to 0.048 and 0.91 for the malignant lesions. The scattering parameters (δ_2 and β_2) for the bottom layer showed minimal differences between the normal/benign tissue and malignant lesions. It should be noted that the larger the value of β, the higher the contribution of the smaller size particles in the scattering particle size distribution function. Thus, an increase in the β value indicates a decrease in the scattering particle average size (Fawzy and Zeng 2006; Schmitt and Kumar 1998). Statistical analysis, using the Kolmogorov–Smirnov two-sample test, showed that the malignant group has a significant increase in the blood volume fraction, ρ ($p = 0.001 < 0.05$), significant decrease in the oxygen saturation parameter, α ($p = 0.022 < 0.05$), and significant decrease in the mucosa layer scattering volume fraction, δ_1 ($p = 0.013 < 0.05$) compared to the benign group. The results also showed a moderate significant decrease in the size-distribution parameter of the mucosa layer (β_1) in the malignant group compared to the benign group ($p = 0.095 < 0.1$).

The significant increase in the blood volume fraction of the malignant lesions measured in our study agreed with the biological observations that tumors and cancerous tissues exhibit increased microvasculature and accordingly increased blood content (Jain 1988). The significant decrease in the blood oxygenation in the malignant lesions is consistent with the hypoxia-related changes occurring during cancerous development (Höckel and Vaupel 2001), which could be related to the increase in

TABLE 6.1

The Mean and Standard Deviation of the Six Parameters Related to Tissue Absorption and Scattering for the Normal/Benign Tissue Group and the Malignant Lesion Group. Also Shown are the Statistical Test Results on the Significance of the Differences Between the Two Groups for These Six Parameters.

| Parameter | Normal/Benign | | Malignant | | Significance |
	Mean	Std. Dev.	Mean	Std. Dev.	(p)
ρ	0.032	0.02	0.065	0.03	0.001
α	0.90	0.11	0.78	0.13	0.022
δ_1	0.077	0.057	0.048	0.046	0.013
β_1	0.97	0.15	0.91	0.12	0.095
δ_2	0.066	0.048	0.07	0.032	0.25
β_2	0.94	0.12	0.92	0.1	0.65

tissue metabolism rate, the lower quality of the tumoral microcirculation, and to the high proliferation rate of the cancerous cells. The significant decrease in the scattering volume fraction found in the measured malignant lesions is consistent with the results obtained by Bard et al. (2005b) for the lung cancer lesions and that obtained by Zonios et al. 1998 for the colon polyps. The mechanism for such a decrease in the scattering volume fraction is still poorly understood due to the complex nature of the tissue scattering process. However, this may be related to the decrease in the mitochondrial content in the cell nucleus (Beauvoit and Chance 1998), which have been found to contribute most significantly to the light scattering in the backward (reflectance) directions (Mourant et al. 2001) or to the changes in the refractive index of the cytoplasm due to an increased protein and enzyme content. The size-distribution parameter (β_1) decrease means increased scatterer particle sizes on average for malignant tissues as compared to normal/benign tissues. This is consistent with the fact that cancerous cells have a larger nucleus than normal and benign cells (Gillaud et al. 2005).

The results obtained from the DFA showed that the three parameters (ρ, α, δ_1) were significant for the discrimination between the two groups. Figure 6.10a shows the classification results based on measuring the blood volume fraction (ρ) and the scattering volume fraction (δ_1) and Figure 6.10b shows the classification results based on measuring the blood volume fraction and the tissue oxygen saturation parameter. As shown in the figure we can easily identify two domain spaces, with slight overlap, for benign and malignant groups. The DFA results using the three parameters (ρ, α, δ_1) with the leave-one-out, cross-validation method showed that we could differentiate the measured lesions into normal/benign and malignant with sensitivity and specificity of 83% and 81%, respectively. In comparison, we also calculated the sensitivity and specificity of imaging (WL + FL) diagnosis for the same patient population to be 87% and 43%, respectively. The relative improvement on detection

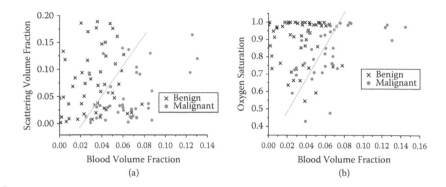

FIGURE 6.10   Plot of (a) blood volume fraction ρ versus the scattering volume δ_1, (b) blood volume fraction ρ versus oxygen saturation parameter α. (Reproduced with permission from Fawzy, Y.S. et al. 2006. *J. Biomed. Opt.* 11: 044003[1–11].)

specificity for spectral diagnosis over imaging diagnosis is obvious, and the magnitude of improvement is significant. It should also be noted that the imaging diagnosis is a somewhat subjective procedure that depends on the experience of the attending physician on using fluorescence endoscopy, while the spectral diagnosis presented in this study is a quantitative and objective method.

REFERENCES

Amelink, A., Sterenborg, H.J.C.M., Bard, M.P.L., and Burgers, S.A. 2004. *In vivo* measurement of the local optical properties of tissue by use of differential path-length spectroscopy. *Opt. Lett.* 29: 1087–1089.

Bard, M.P.L., Amelink, A., Hegt, V.N., Graveland, W.J., Sterenborg, H.J.C.M., Hoogsteden, H.C., and Aerts, J.G.J.V. 2005a. Measurement of hypoxia-related parameters in bronchial mucosa by use of optical spectroscopy. *Am. J. Respir. Crit. Care Med.* 171: 1178–1184.

Bard, M.P.L., Amelink, A., Skurichina, M., den Bakkerd, M., Burgers, S.A., van Meerbeeck, J.P., Duin, R.P.W., Aerts, J.G.J.V., Hoogsteden, H.C., and Sterenborg, H.J.C.M. 2005b. Improving the specificity of fluorescence bronchoscopy for the analysis of neoplastic lesions of the bronchial tree by combination with optical spectroscopy: Preliminary communication. *Lung Cancer* 47: 41–47.

Beauvoit B. and Chance B. 1998. Time-resolved spectroscopy of mitochondria, cells and tissue under normal and pathological conditions. *Mol. Cell Biochem.* 184: 445–455.

Bohren, C.F. and Huffman, D.R. 1983. *Absorption and Scattering of Light by Small Particles*. New York, John Wiley & Sons.

Dillion, R.W. and Goldstein, M. 1984. *Multivariate Analysis: Methods and Applications*. New York: John Wiley & Sons.

Fawzy, Y.S. and Zeng, H. 2006. Determination of scattering volume fraction and particle size-distribution in the superficial layer of a turbid medium using diffuse reflectance spectroscopy. *Appl. Opt.* 45: 3902–3912.

Fawzy, Y.S., Petek, M., Tercelj, M., and Zeng, H. 2006. *In-vivo* assessment and evaluation of lung tissue morphologic and physiological changes from non-contact endoscopic reflectance spectroscopy for improving lung cancer detection, *J. Biomed. Opt.* 11: 044003(1–11).

Fawzy, Y. and Zeng, H. 2008. Intrinsic fluorescence spectroscopy for endoscopic detection and localization of the endo-bronchial cancerous lesions. *J. Biomed. Opt.* 13: 064022(1–8).

Gelebart, B., Tinet, E., Tualle, J.M., and Avrillier, S. 1996. Phase function simulation in tissue phantoms: A fractal approach. *Pure Appl. Opt.* 5: 377–388.

Gillaud M., le Riche J. C., Dawe C., Korbelik J, Coldman A., Wistuba I. I., Park I., Gazdar A., Lam S., and MacAulay C. E. 2005. Nuclear morphometry as a biomarker for bronchial intraepithelial neoplasia: Correlation with genetic damage and cancer development. *Cytometry Part A* 63A: 34–40.

Höckel, M. and Vaupel, P. 2001. Tumor hypoxia: Definitions and current clinical biologic, and molecular aspects. *J. Natl. Cancer Inst.* 93: 266–276.

Hung, J., Lam, S., LeRiche, J., and Palcic, B. 1991. Autofluorescence of normal and malignant bronchial tissue. *Laser Surg. Med.* 11: 99–105.

Jain, R.K.1988. Determinants of tumor blood flow: A review. *Cancer Res.* 48: 2641–2658.

Lachenbruch, P. and Mickey, R.M. 1968. Estimation of error rates in discriminant analysis. *Technometrics* 10: 1–11.

Mourant, J.R., Johnson, T.M., and Freyer, J.P. 2001. Characterizing mammalian cells and cell phantoms by polarized backscattering fiber-optic measurement. *Appl. Opt.* 40: 5114–5123.

Press, W.H., Teukolsky, S.A., Vetterling, W.T., and Flannery, B.P. 1992. *Numerical Recipes in C*, 2nd ed. Cambridge, UK: Cambridge University Press.

Qu, J., MacAulay, C., Lam, S., and Paclic, B. 1994. Optical properties of normal and carcinomatous bronchial tissue. *Appl. Opt.* 31: 7397–7405.

Schmitt, J.M. and Kumar, G. 1996. Turbulent nature of refractive-index variations in biological tissue. *Opt. Lett.* 21: 1310–1312.

Schmitt, J.M. and Kumar, G. 1998. Optical scattering properties of soft tissue: A discrete particle model. *Appl. Opt.* 37: 2788–2797.

Utzinger, U. and Richards-Kortum, R.R. 2003. Fiber optic probes for biomedical optical spectroscopy. *J. Biomed. Opt.* 8: 121–147.

Venugopalan, V., You, J.S., and Tromberg, B.J. 1998. Radiative transport in the diffusion approximation: An extension for highly absorbing media and small source-detector separation. *Phys. Rev. E* 58: 2395–2407.

Zeng, H., MacAulay, C., Palcic, B., and McLean, D.I. 1993. A computerized autofluorescence and diffuse reflectance spectroanalyser system for *in vivo* skin studies. *Phys. Med. Biol.* 38: 231–240.

Zeng, H., MacAulay, C., McLean, D.I., Lui, H., and Palcic, B. 1995. Miniature spectrometer and multi-spectral imager as a potential diagnostic aid for dermatology. *SPIE Proc.* 2387: 57–61.

Zeng, H., Petek, M., Zorman, M.T., McWilliams, A., Palcic, B., and Lam, S. 2004. Integrated endoscopy system for simultaneous imaging and spectroscopy for early lung cancer detection. *Opt. Lett.* 29: 587–589.

Zonios, G., Perelman, L.T., Backman, V., Manoharan, R., Fitzmaurice, M., Van Dam, J., and Feld, M.S. 1998. Diffuse reflectance spectroscopy of human adenomatous colon polyps *in vivo*. *Appl. Opt.* 38: 6628–6636.

7 Endoscopic Raman Spectroscopy

Haishan Zeng
British Columbia Cancer Agency Research Centre,
Vancouver, British Columbia, Canada

CONTENTS

When monochromatic light strikes a sample, almost all the observed light is scattered elastically (Rayleigh scattering) with no change in energy (or frequency). A very small portion of the scattered light, ~1 in 10^{10}, is inelastically scattered (Raman scattering) with a corresponding change in frequency. The difference between the incident and scattered frequencies corresponds to an excitation of the molecular system, most often excitation of vibrational modes. By measuring the intensity of the scattered photons as a function of the frequency difference, a Raman spectrum is obtained. Different from fluorescence spectrum and reflectance spectrum, Raman peaks are typically narrow and in many cases can be attributed to the vibration of specific chemical bonds (or normal mode dominated by the vibration of a functional group) in a molecule. As such, it is a "fingerprint" for the presence of various molecular species and can be used for both qualitative identification and quantitative determination (Grasselli and Bulkin 1991).

In the 1980s and 1990s, Raman spectroscopy study of various biological tissues including brain, breast, bladder, colon, larynx, cervix, lung, and skin, etc., had been reported (see, for example, Ozaki 1988; Liu et al. 1992; Mahadevan-Jansen and Richards-Kortum 1996; Lawson et al. 1997). Identified Raman scatters in tissues include elastin, collagen, blood, lipid, tryptophan, tyrosine, carotenoid, myoglobin, nucleic acids, etc. Most early studies obtained data from *ex vivo* tissue samples using Fourier-transform (FT) Raman spectrometers. These data have demonstrated the great potential of Raman spectroscopy for disease diagnosis. For clinical applications,

Raman measurements must be performed *in vivo* and in real-time (Zeng et al. 2008). This is challenging because Raman signals are extremely weak. Single-channel laboratory Fourier-transform (FT) Raman systems add many scans to obtain spectra with good signal-to-noise ratios. This can take up to 30 min and, therefore, has limited clinical utility. The first major advancement toward real-time *in vivo* Raman spectroscopy was reported by Baraga et al. (1992). The instrument was a dispersive spectrograph with a CCD multichannel near-infrared (NIR) detector. In contrast to FT-Raman systems, this instrument simultaneously detected photons of different wavelengths to obtain spectra from excised human aortic tissue with short, 5 min integration times. Since then, optimized dispersive-type Raman instruments based on fiber-optic light delivery and collection, compact diode lasers, and highly efficient spectrograph-detector combinations have been reported (e.g., Mahadevan–Jansen et al. 1998; Shim et al. 2000). These advances have lead to increased sensitivity and decreased *in vivo* measurement time. Some of them are capable of acquiring an *in vivo* Raman spectra within seconds or subseconds through the endoscope, paving the way for clinical applications (Huang et al. 2001; Motz et al. 2004; Short et al. 2008; Huang et al. 2009; Short et al. 2011; Zeng et al. 2011). In this chapter, we will summarize our technology development of a rapid Raman system and its application for lung cancer detection.

7.1 RAPID RAMAN SYSTEM

7.1.1 System Configuration

In 2001, we developed a rapid Raman system for *in vivo* skin applications (Huang et al. 2001; Zhao et al. 2008), which later on also became the engine for our endoscopy Raman system (Short et al. 2008, 2011). Figure 7.1 shows the block diagram of the system. It consisted of a stabilized diode laser (785 nm), transmissive imaging spectrograph, CCD detector (NIR-optimized and deep depletion back-illuminated), and a specially-designed Raman probe. For rapid spectral acquisition, the spectrograph uses a volume phase technology holographic grating for high-throughput light dispersion to facilitate simultaneous whole-spectrum detection by the CCD array detector. We designed the probe to maximize the collection of Raman signals from the tissues while reducing interfering fluorescence and Raman signals from the fibers. Figure 7.2 shows the probe, which consists of illumination and Raman collection arms. To enhance signal detection, we packed as many fibers into the collection arm as permitted by the CCD height (6.9 mm). The fiber bundle consisted of 58 fibers (each 100 μm in diameter) arranged in a circle at the input end and a linear array at the output end, which is connected to the spectrograph. We arranged the fibers of the linear array along a parabolic curve to correct for spectrograph image aberration (see Figure 7.3). This also allowed hardware binning of the vertical spectral line on the CCD. Hardware binning here is the capability of the CCD to accumulate electrons from different pixels in one vertical column (i.e., channel) that correspond to the same light wavelength. This proprietary technology development preserved spectral resolution and yielded a 16-fold improvement in the signal-to-noise ratio. Using our

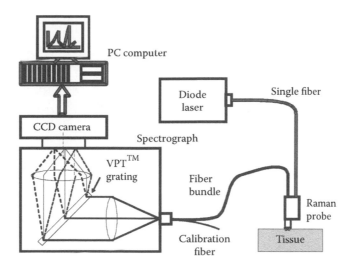

FIGURE 7.1   Block diagram of the rapid, dispersive Raman spectrometer system. The spectrograph uses a volume phase technology (**VPT**) holographic grating as the dispersing element. A back-illuminated CCD camera detects photons of different wavelengths simultaneously for rapid spectral acquisition. (Reproduced with permission from Huang, Z. et al. 2003. *Intl. J. Cancer* 107: 1047–1052.)

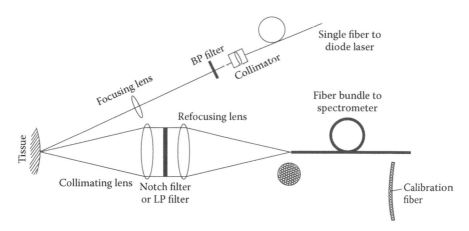

FIGURE 7.2   Optical layout of the skin Raman probe with a band pass (BP) filter in the illumination arm and a circular-to-parabolic fiber bundle in the collection arm. (Reproduced with permission from Huang, Z. et al. 2001. *Opt. Lett.* 26: 1782–1784.)

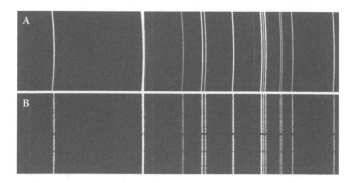

FIGURE 7.3 (A) Image of a slit (100 μm) on the CCD through the spectrograph showing image aberration. (B) CCD image of 58 fibers aligned along a parabolic curve at the entrance of the spectrograph, where the image aberration has been corrected. (Reproduced with permission from Huang, Z. et al. 2001. *Opt. Lett.* 26: 1782–1784.)

system, a tissue Raman spectrum can be acquired within 1 s, a vast improvement compared to 20–90 s acquisition time for other systems at the time.

7.1.2 System Calibration

The system wavelength calibration was performed with Ne and Hg standard lamps. With a LF grating, the system covered the low-frequency fingerprint spectral range up to 1800 cm^{-1}. While using a HF grating, the system is capable of measuring Raman spectra from 1500 cm^{-1} to 3500 cm^{-1}. Raman frequencies were calibrated with cyclohexane, acetone, and barium sulfate to an accuracy of ±2 cm^{-1}, and the spectral resolution was 8 cm^{-1}. All spectra acquired were corrected for the spectral sensitivity of the system by use of a standard lamp (RS-10, EG&G Gamma Scientific).

7.1.3 Data Processing

One of the major challenges for biomedical Raman spectroscopy is the removal of intrinsic autofluorescence background signals, which are usually a few orders of magnitude stronger than those arising from Raman scattering. A number of methods have been proposed for fluorescence background removal including excitation wavelength shifting, Fourier transformation, time gating, and simple or modified polynomial fitting. The single polynomial and the modified multipolynomial fitting methods are relatively simple and effective, and thus widely used in biological applications. However, their performance in real-time *in vivo* applications and low signal-to-noise ratio environments is suboptimal. We have developed an improved automated algorithm (Zhao et al. 2007) for fluorescence removal based on modified multipolynomial fitting, but with the addition of (1) a peak-removal procedure during the first iteration, and (2) a statistical method to account for signal noise effects. Experimental results demonstrate that this approach improves the automated removal of the fluorescence background during real-time Raman spectroscopy and for *in vivo* measurements characterized by low signal-to-noise ratios.

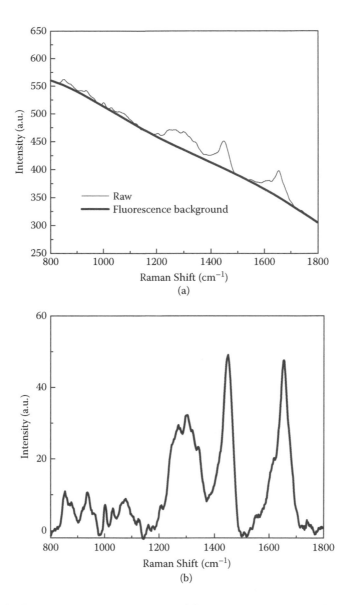

FIGURE 7.4 Fluorescence background removal from a volunteer's palm skin Raman measurement. (Reproduced with permission from Zeng, H. et al. 2008. *J. Innovative Opt. Health Sci.* 1: 95–106.)

Figure 7.4a shows the raw data and the fitted fluorescence background, while Figure 7.4b shows the pure Raman signal after subtracting the fitted fluorescence background from the raw data. Details of this algorithm can be found in Zhao et al. 2007 and a computer code can be downloaded from http://www.bccrc.ca/dept/ic/cancer-imaging/haishan-zeng-phd.

7.1.4 Study of Excised Lung Tissues

With the above system, we studied Raman spectra of fresh bronchial tissue samples obtained from biopsy or surgical resection (Huang et al. 2003). Figure 7.5 shows the normalized mean Raman spectra of normal and cancerous bronchial tissues. It can be seen that significant Raman spectral differences exist between normal and tumor tissues, while Raman spectra of adenocarcinoma and squamous cell carcinoma are very similar to each other. The ratio of Raman intensities at 1445 cm^{-1} to 1655 cm^{-1} provided good differentiation between normal and tumor tissues. Tumors showed a higher percentage of Raman signals for nucleic acid, tryptophan, and phenylalanine, but lower percentage signals for phospholipids, praline, and valine, compared to normal tissue (Figure 7.5). Table 7.1 shows a tentative assignment of major Raman bands observed from the spectra. A variety of biochemical information can be obtained. The results of this exploratory study indicate that NIR Raman spectroscopy may have significant potential for improving lung cancer diagnosis based on optical evaluation of biomolecules. The development of an endoscopic Raman probe for *in vivo* Raman spectral measurements is encouraged.

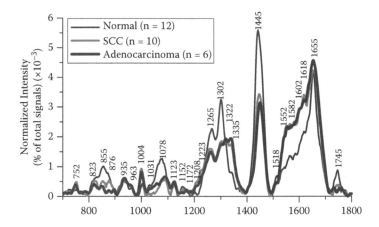

FIGURE 7.5   The mean Raman spectra of normal bronchial tissue (n = 12), adenocarcinoma (n = 6) and squamous cell carcinoma (n = 10). Each spectrum was normalized to the integrated area under the curve to correct for variations in absolute spectral intensity. Each spectral peak can be assigned to specific biomolecules inside the tissue. The Raman signatures for tumor are significantly different from normal tissue. (Reproduced with permission from Huang, Z. et al. 2003. *Intl. J. Cancer* 107: 1047–1052.)

TABLE 7.1

Peak Positions and Tentative Assignments of Major Vibrational Bands Observed in Normal and Tumor Bronchial Tissue

Peak Position (cm⁻¹)	Protein Assignments	Lipid Assignments	Others
1745 w		v (C = O), phospholipids	
1655 vs	v (C = O) amide I, α-helix, collagen, elastin		
1618 s (sh)	v (C = C), tryptophan		N (C = C), porphyrin
1602 ms (sh)	δ (C = C), phenylalanine		
1582 ms (sh)	δ (C = C), phenylalanine		
1552 ms (sh)	v (C = C), tryptophan		v (C = C), porphyrin
1518 w			v (C = C), carotenoid
1445 vs	δ (CH₂), δ (CH₃), collagen	δ (CH₂) scissoring, phospholipids	
1335 s (sh)	CH₃CH₂ wagging, collagen		CH₃CH₂ wagging nucleic acids
1322 s	CH₃CH₂ twisting, collagen		
1302 vs	δ (CH₂) twisting, wagging, collagen	δ (CH₂) twisting, wagging, phospholipids	
1265s (sh)	v (CN), δ (NH) amide III, α-helix, collagen, tryptophan		
1223 mw (sh)			v_{as}(PO₂), nucleic acids
1208 w (sh)	v (C-C₆H₅), tryptophan, phenylalanine		
1172 vw	δ (C-H), tyrosine		
1152 w	v (C-N), proteins		v (C-C), carotenoid
1123 w	v (C-N), proteins		
1078 ms		v (C-C) or v (C-O), phospholipids	
1031 mw (sh)	δ (C-H), phenylalanine		
1004 ms	v_s (C-C), symmetric ring breathing, phenylalanine		
963 w	Unassigned		
935 w	v (C-C), α-helix, proline, valine		
876 w (sh)	v (C-C), hydroxyproline		
855 ms	v (C-C), praline		Polysaccharide
	δ (CCH) ring breathing, tyrosine		
823 w	out-of-plane ring breathing, tyrosine		
752 w	symmetric breathing, tryptophan		

Source: Huang, Z. et al. 2003. *Intl. J. Cancer* 107: 1047–1052. With permission.

Note: v is for stretching mode; v_s, symmetric stretch; v_{as}, asymmetric stretch; δ, bending mode; v = very; s = strong; m = medium; w = weak; sh = shoulder.

7.2 CATHETER DEVELOPMENT FOR ENDOSCOPIC RAMAN SPECTROSCOPY

We have successfully developed an endoscopic laser-Raman catheter for real-time *in vivo* lung Raman measurements (Short et al. 2008). The key specifications of the endoscopic laser-Raman catheter are (1) small enough to pass through the broncho-scope instrument channel (2.2 mm size), (2) incorporating proper filtering mech-anism to minimize or eliminate the background Raman and fluorescence signals generated from the fiber itself, and (3) able to collect enough signal so that a Raman spectrum can be acquired in seconds or subseconds. To preserve the high S/N ratio advantage of our skin Raman probe, we employed a two-step filtering strategy for the endoscopic Raman catheter: (1) first-order filtering at the tip of the fiber bundle and (2) high performance filtering at the entrance point of the instrument channel of the endoscope. Figure 7.6 shows the schematics of the catheter we built. It consists of a probing fiber bundle assembly, a filter adapter, an illumination fiber, and a round to parabolic linear array fiber bundle. The 785 nm laser light is focused into the illumi-nation fiber, which is connected to the filter adapter close to the instrument channel entrance. A high performance BP (band pass) filter (785 ± 2.5 nm) passes through the laser light and filters out the background Raman and fluorescence signals gener-ated inside the illumination fiber between the laser and the filter adapter. The filtered laser light is refocused into the illumination fiber in the probing fiber bundle assem-bly. Because this part of the illumination fiber is short, the generated background Raman and fluorescence is small. Nevertheless, the distal end of the illumination fiber is coated with a SP (short pass) filter to further reduce these background signals. The induced Raman signal from the tissue is picked up by collection fibers in the

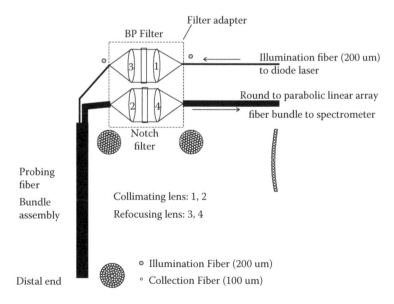

FIGURE 7.6 Schematic diagram of the endoscopic laser-Raman catheter. (Reproduced with permission from Zeng, H. et al. 2008. *J. Innovative Opt. Health Sci.* 1: 95–106.)

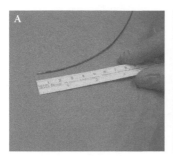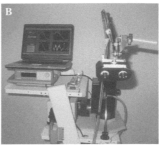

FIGURE 7.7 Photographs of the endoscopic laser-Raman system. (A) The distal end of the Raman catheter (1.8 mm diameter). (B) The complete Raman system mounted on a movable cart. (Reproduced with permission from Short, M. et al. H. 2011. *J. Thoracic Oncol.* 6: 1206–1214.)

probing fiber bundle assembly. LP filter coatings are applied to these fibers to block the back-scattered laser light from entering the probe. At the proximal end of the probing fiber bundle assembly, these collection fibers are packed into a round bundle and connected to the filter adapter. A notch filter (OD > 6 at 785 nm, Kaiser) is used to further block the laser wavelengths and allow the Raman signals to pass through. The Raman signals are refocused by lens 4 into the round-to-parabolic linear array fiber bundle. At the entrance of the spectrometer, these collection fibers are aligned along a parabolic line to correct for image aberration of the spectrograph to achieve better spectral resolution and higher S/N ratio in a fashion similar to our skin Raman probe (Figure 7.2). In the lung Raman system, the endoscopic laser-Raman catheter replaces the skin Raman probe of the skin Raman system as shown in Figures 7.1 and 7.2. Figure 7.7 is a selection of photographs showing the Raman catheter and the completed endoscopy Raman system.

We tested the system and the Raman catheter on lung patients. Initial measurements in the low frequency (LF) range (700 cm^{-1} to 1800 cm^{-1}) resulted in spectra dominated by a fluorescence emission. Obtaining precise Raman spectra under such a high fluorescence background was difficult. This is an example that *ex vivo* study results (Figure 7.5) cannot always be translated into *in vivo* clinical applications. We thus shifted to measure the Raman spectra in the high frequency (HF) range (1500 cm^{-1} to 3500 cm^{-1}). And it worked well with tissue autofluorescence significantly reduced. Figure 7.8 shows examples of HF *in vivo* spectra of lung sites with different pathologies. Clear Raman peaks are seen in the raw data (A). The fluorescence intensities were much less than in the LF range. Polynomial fitting for removing the fluorescence results in nice Raman spectra shown in (B). The drawback as compared to LF Raman was that there were fewer peaks in the HF range. The dashed lines shown in Figure 7.8 indicate the position of most prominent peaks. The broad peaks near 1658 cm^{-1}, were probably due to a combination of v (C=O) amide I vibrations (Huang et al. 2003) and v_2 water molecule bending motions (Chaplin 2013; Movasaghi et al. 2007). The intense peaks at 2900 cm^{-1} are assigned to a combination of CH_2 antisymmetric (2880 cm^{-1}) and CH_3 symmetric (2935 cm^{-1}) stretching modes of phospholipids. The broad peaks near 3300 cm^{-1} were most likely due

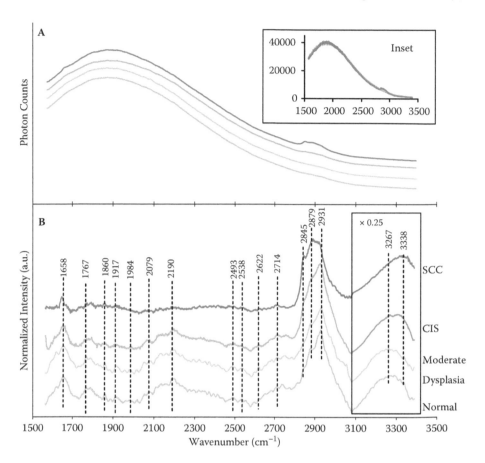

FIGURE 7.8   *In vivo* spectra of human lung tissue with different pathologies in the HF range (2 s integration). (A) main graph: smoothed raw data (shifted along y axis for clarity). Top to bottom order: SCC, CIS, moderate dysplasia, and normal tissue. (A) inset: unprocessed raw data. (B) Extracted Raman spectra (intensity calibrated, normalized, and shifted along y axis for clarity) in same top to bottom order as (A) on main graph. The data above 3100 cm^{-1} has been scaled by 0.25. (Reproduced with permission from Short, M. et al. 2008. *Opt. Lett.* 33: 711–713.)

to overlapping v_1 and v_3 symmetric and antisymmetric OH stretching motions of water. Between 1658 and 2885 cm^{-1} there are a number of low intensity peaks that show considerable variability between the spectra. Generally, this spectral region is not noted for any significant Raman peaks. However, a comprehensive review by Movasaghi et al. (2007) lists a number of possibilities, mainly involving carbon, nitrogen, and oxygen modes which are in approximate agreement with many of the peaks we observe. The squamous cell carcinoma (SCC) spectrum was dramatically different from the other three spectra, and the moderate dysplasia spectrum was very close to the normal spectrum. The carcinoma *in situ* (CIS) spectrum had features similar to the SCC spectrum, but also features similar to the moderate dysplasia spectrum. Overall, the spectra we obtained from the two malignant lesions (SCC and CIS) could be easily differentiated from the spectra we obtained from the benign lesion (moderate dysplasia) and normal tissue.

7.3 COMBINED RAMAN SPECTROSCOPY AND FLUORESCENCE ENDOSCOPY FOR IMPROVING LUNG CANCER DETECTION

Fluorescence endoscopy has significantly improved the diagnostic sensitivity and is currently the most sensitive field imaging modality for localizing lung cancers. However, its specificity is about 60%, equivalent to a 40% false positive biopsy rate (for references, see Chapter 12). Therefore, it is highly desirable to develop point spectroscopy or "point" microscopy to be used in combination with fluorescence endoscopy to improve the specificity. Using the Raman system and the Raman catheter described in the last two sections (7.1 and 7.2), we have conducted a pilot study to test if endoscopy Raman spectroscopy in combination with fluorescence endoscopy can improve lung cancer detection, especially the diagnostic specificity.

The procedure was for the bronchoscopist to identify lesions they would normally biopsy using combined white light imaging and fluorescence imaging (see Figure 7.9A,B). Raman spectra were obtained from these sites (see Figure 7.9C). Biopsies were taken of the same locations, and classified by a pathologist. Eight classifications were used according to WHO criteria (Travis et al. 1999): Normal epithelium, Hyperplasia, Metaplasia, Mild Dysplasia, Moderate Dysplasia, Severe Dysplasia, CIS, and Invasive Squamous Cell Carcinoma (IC). The presence or absence of inflammatory changes was also recorded. A total of 46 patients participated in the study; of these 26 were found to have suspicious areas with white light and fluorescence imaging. There were 129 Raman spectra obtained; 51 of these were from sites with pathologies of ≥MOD (moderate dysplasia), the rest were from sites with pathologies of mild dysplasia or better (≤MILD).

The measured raw Raman spectra (contain both the fluorescence background and the Raman signals) were preprocessed into three different datasets for subsequent

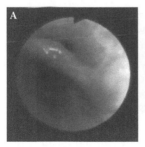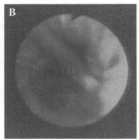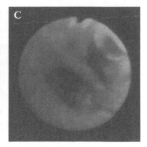

FIGURE 7.9 (See color insert) Pictures showing images acquired by a multimode bronchoscopy system. (A) white light image, and (B) blue light excited fluorescence image (Xillix, Onco-LIFE) of the same location as (A). For image (B) green is representative of normal tissue, and dark red (centre of image) is diseased tissue. (C) same lesion being excited simultaneously with blue light to generate a fluorescence image and with 785 nm light from the Raman catheter, which can be seen in the top right corner of the image. (Reproduced with permission from Short, M. et al. H. 2011. *J. Thoracic Oncol.* 6: 1206–1214.)

data analysis. In the first case (dataset A) each spectrum was smoothed (with a 3-point moving average) to remove random noise, and then normalized to reduce the effect of intensity variations from different tissue sites. The normalization was accomplished by summing the area under each curve and dividing each variable in the smoothed spectrum by this sum. In the second case (dataset B) each spectrum was also smoothed, but in this case the autofluorescence was subtracted by a modified polynomial-fitting routine before normalization (Zhao et al. 2007). The third method (dataset C) used a Savitzky–Golay six point quadratic polynomial on each spectrum to calculate a smoothed second-order derivative spectrum (Savitzky and Golay 1964). Summing the squared derivative values of a spectrum and then dividing each variable by this sum was used for normalization.

Datasets A, B, and C were analyzed separately using statistical software (Statistica 6.0, StatSoft Inc. Tulsa, OK). Multivariate statistics were used which compare whole spectra rather than single Raman peaks (univariate) as these have been shown to be more accurate (Cooper 1999). Principle components (PCs) for all the spectra in each dataset were computed to reduce the number of variables. Student's t-tests were used on PCs that accounted for 0.1% or more of the variance to determine those most significant at separating spectra into two pathology groups: $\geq$MOD and $\leq$MILD. A linear discrimination analysis (LDA) with leave-one-out cross validation was used on the most significant PCs. To avoid overfitting the data, the number of PCs used in the LDA were limited to one-third (17) of the total number of cases of the smallest subgroup, that is, 51 $\geq$MOD spectra.

The statistical analyses on spectra from datasets A, B, and C led to significantly different results. Spectra from dataset A were the worst in predicting the pathology $\geq$MOD with 80% sensitivity and 72% specificity. Analyses of dataset B spectra showed an improvement in pathology prediction compared to dataset A spectra with 80% sensitivity and 79% specificity. The best result was obtained by analyzing spectra processed with the second-order derivative (dataset C). 90% sensitivity and 91% specificity were obtained. The second derivative spectra of dataset C, over the ranges where significant differences ($p \leq 0.05$) between different pathology groups were apparent, are shown in Figure 7.10. The posterior probability plot generated by the classification algorithm is shown in Figure 7.11. It shows that only three IC spectra were misclassified. If we drop all the IC spectra from analyses, the resulting sensitivity increases to 96% with the specificity unchanged at 91%. The receiver operating characteristics (ROC) for all the three datasets are shown in Figure 7.12. One can clearly see the superiority of second-order derivative-processed spectra (dataset C). The integrated areas under each ROC curve were 0.78, 0.85, and 0.92 for spectra analyzed in datasets A, B, and C, respectively.

In conclusion, it appears that point Raman spectroscopy can significantly reduce the number of false positive biopsies, while marginally reducing the sensitivity of white light and fluorescence imaging to the detection of preneoplastic lung lesions. Although it may be considered better to have a 40% false positive rate than incur any loss in detection sensitivity, the slight loss incurred with the adjunct use of Raman may not be realized in practice. Currently, bronchoscopists have to make partially subjective decisions when using white light and fluorescence imaging about which lesions to biopsy. The adjunct use of Raman would make the decision process more

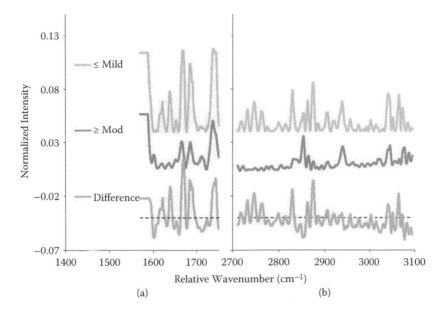

FIGURE 7.10 Average spectra of dataset C processed data from sites with pathology ≤MILD (upper curve) and ≥MOD (middle curve). The two averages have been shifted on the intensity scale for clarity. The lower plot shows the result of subtracting the average ≤MILD spectra from the average ≥MOD spectra (not on the same intensity scale). The horizontal dashed line is at zero intensity. Two ranges (*a* and *b*) are shown as these were only where clear Raman peaks were observed. (Reproduced with permission from Short, M. et al. H. 2011. *J. Thoracic Oncol.* 6: 1206–1214.)

objective, which may result in the identification of additional preneoplastic lesions at sites initially rejected as biopsy candidates by imaging. Thus, the sensitivity could increase. This study suggests that Raman spectroscopy in combination with white light and fluorescence imaging have great potential for improving the early detection of preneoplastic lesions.

It should be pointed out that the commercialization of Raman spectroscopy for endoscopy application has been initiated by a Canadian company (Verisante Technology, Inc.). They have successfully launched a regulatory approved system for skin cancer detection using a Raman spectroscopy-based algorithm (Lui et al. 2012). Spectral data acquisition, Raman algorithm analysis, and diagnostic results' display are all completed in less than 2 s. It is anticipated that a similar performance will be realized for an endoscopy Raman diagnostic device.

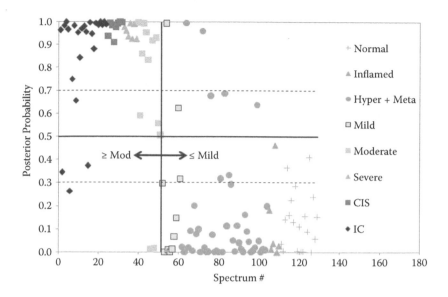

FIGURE 7.11 Posterior probability plot of predicted pathology compared with known pathology. Statistical analyses were performed on dataset C data (second-order derivative spectra) gathered into spectra from ≤MILD and ≥MOD tissue sites using a leave-one-out cross validation. (Reproduced with permission from Short, M. et al. H. 2011. *J. Thoracic Oncol.* 6: 1206–1214.)

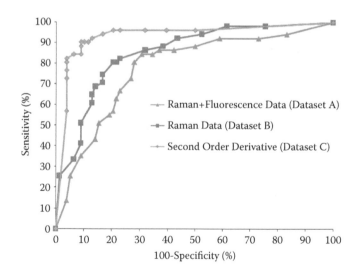

FIGURE 7.12 ROC curves showing how the sensitivity and specificity change when moving the cut line from 0 to 100% in the LDA posterior probability plots. Three curves are shown which belong to data that was statistically analyzed after processing into datasets A, B, and C, respectively. LDA, linear discrimination analysis. (Reproduced with permission from Short, M. et al. H. 2011. *J. Thoracic Oncol.* 6: 1206–1214.)

REFERENCES

Baraga, J. J., Feld, M. S., and Rava, R. P. 1992. Rapid near-infrared Raman spectroscopy of human tissue with a spectrograph and CCD detector. *Appl. Spectrosc.* 46: 187–190.

Chaplin, M. 2013. "Water Structure and Science," http://www.lsbu.ac.uk/water/vibrat.html.

Cooper, J. B. 1999. Chemometric analysis of Raman spectroscopic data for process control applications. *Chemom. Intell. Lab. Syst.* 46: 231–247.

Grasselli, J. G. and Bulkin, B. J. 1991. *Analytical Raman Spectroscopy, Chemical Analysis Series Vol. 114*. New York: John Wiley & Sons.

Huang, Z., Zeng, H., Hamzavi, I., McLean, D. I. and Lui, H. 2001. A rapid near-infrared Raman spectroscopy system for real-time *in vivo* skin measurements. *Opt. Lett.* 26: 1782–1784.

Huang, Z., McWilliams, A., Lui, H., McLean, D. I., Lam, S., Zeng, H. 2003. Near-infrared Raman spectroscopy for optical diagnosis of lung cancer, *Intl. J. Cancer.* 107: 1047–1052.

Huang, Z., Teh, S. K., Zheng, W., Mo, J., Lin, K., Shao, X., Ho, K. Y., Teh, M., and Yeoh, K. G. 2009. Integrated Raman spectroscopy and trimodal wide-field imaging techniques for real-time *in vivo* tissue Raman measurements at endoscopy. *Opt. Lett.* 34: 758–760.

Lawson, E. E., Barry B. W., Williams A. C., and Edwards H. G. M. 1997. Biomedical applications of Raman spectroscopy. *J. Raman Spectrosc.* 28: 111–117.

Liu, C. H., Das, B. B., Glassman, W. L., Tang, G. C., Yoo, K. M., Zhu, H. R., Akins, D. L., Lubicz, S. S., Cleary, J., Prudente, R., Celmer, E., Caron, A., and Alfano, R. R. 1992. Raman, fluorescence and time-resolved light scattering as optical diagnostic techniques to separate diseased and normal biomedical media. *J. Photochem. Photobiol. B: Biol.* 16: 187–209.

Lui, H., Zhao, J., Mclean, D. I., and Zeng, H. 2012. Real-time Raman spectroscopy for *in vivo* skin cancer diagnosis. *Cancer Res.* 71: 2491–2500.

Mahadevan-Jansen, A. and Richards-Kortum, R. 1996. Raman spectroscopy for the detection of cancers and precancers. *J. Biomed. Opt.* 1: 31–70.

Mahadevan-Jansen, A., Mitchell, M. F., Ramanujam, N., Utzinger, U., and Richards-Kortum, R. 1998. Development of a fiber optic probe to measure NIR Raman spectra of cervical tissue *in vivo. Photochem. Photobiol.* 68: 427–431.

Motz, J. T., Hunter, M., Galindo, L. H., Gardecki, J. A., Kramer, J. R., Dasari, R. R., and Feld, M. S. 2004. Optical fiber probe for biomedical Raman spectroscopy. *Appl. Opt.* 43: 542–554.

Movasaghi, Z., Rehman, S., and Rehman, I. U. 2007. Raman spectroscopy of biological tissues. *Appl. Spectrosc. Rev.* 42: 493–541.

Ozaki, Y. 1988. Medical application of Raman spectroscopy. *Appl. Spectrosc. Rev.* 24: 259–312.

Savitzky, A. and Golay, A. 1964. Smoothing and differentiation of data by simplified least squares procedure. *Anal. Chem.* 36: 627–1639.

Shim, M. G., Wong, L. K. S., Marcon, N. E., and Wilson, B. C. 2000. *In vivo* near-infrared Raman spectroscopy: Demonstration of feasibility during clinical gastrointestinal endoscopy. *Photochem. Photobiol.* 72: 146–150.

Short, M., Lam, S., McWilliams, A., Zhao, J., Lui, H., and Zeng, H. 2008. Development and preliminary results of an endoscopic Raman probe for potential in-vivo diagnosis of lung cancers. *Opt. Lett.* 33: 711–713.

Short, M., Lam, S., McWilliams, A., Ionescu, D. N., and Zeng, H. 2011. Using laser Raman spectroscopy to reduce false positives of autofluorescence bronchoscopies: A pilot study. *J. Thoracic Oncol.* 6: 1206–1214.

Travis, W. D., Colby, T. V., Corrin, B. et al. 1999. Histologic and graphical text slides for the histological typing of lung and pleural tumors. In World Health Organization Pathology Panel: World Health Organization. *International Histological Classification of Tumors*, 3rd ed. Berlin: Springer-Verlag, p. 5.

Zeng, H., Zhao, J., Short, M., McLean, D., Lam, S., McWilliams, A., and Lui, H. 2008. Raman spectroscopy for *in vivo* tissue analysis and diagnosis—from instrument development to clinical applications. *J. Innovative Opt. Health Sci.* 1: 95–106.

Zeng, H., Lui, H., and McLean, D. I. 2011. Skin cancer detection using *in vivo* Raman spectroscopy. *SPIE Newsroom*, 11 May 2011. DOI: 10.1117/2.1201104.003705.

Zhao, J., Lui, H., McLean, D. I., and Zeng, H. 2007. Automated autofluorescence background subtraction algorithm for biomedical Raman spectroscopy. *Appl. Spectrosc.* 61: 1225–1232.

Zhao, J., Lui, H., McLean, D. I., and Zeng, H. 2008. Integrated real-time Raman system for clinical *in vivo* skin analysis. *Skin Res. Technol.* 14: 484–492.

Section IV

Endoscopic "Point" Microscopy

8 Endoscopic Confocal Microscopy

Pierre Lane and Calum MacAulay
British Columbia Cancer Agency Research Center,
Vancouver, British Columbia, Canada

Michele Follen
Texas Tech University, El Paso, Texas

CONTENTS

8.1   BACKGROUND AND IMPORTANCE

The goal of endoscopy is to decrease the morbidity and mortality of cancer and other chronic diseases in the world. In 2004, the World Health Organization recorded 59 million deaths worldwide with 13 million due to cardiovascular diseases (coronary heart disease and stroke) (WHO 2009b) and 8 million due to cancer (WHO 2009a). As more of the world's population lives past 70 years of age, most deaths are due to these two common chronic diseases. As we seek to decrease morbidity and mortality in the world by developing applicable technologies (keeping cost-effectiveness, local resources, and infrastructure in mind) we will allow these applications to be relevant to high, middle, and low-income countries. Both of these leading causes of death are chronic diseases, that is, they develop over a lifetime (often over 20–30 years), and so there is an almost equal time interval during which interventions are possible (Khanavkar 1998; Palcic 1991; BCCA 2005; Lane 2006; Rosin 2000, 2002; Zhang 1997; Heintzmann 2001).

The screening and diagnostic paradigms for both cardiovascular diseases and cancer are similar. There are risk factors that can be elucidated by patient history, others by physical exam, and still others through interrogation of biomarkers (biological substances found in the serum, blood cells, or tissue). The stages that cancers evolve through as they develop from normal into cancer are illustrated schematically in Figure 8.1. The process begins with normal cells which progress to intraepithelial neoplasia (IEN) and carcinoma *in situ* (CIS), and finally to invasive cancer. The process is somewhat counterintuitive in that initial developments focuses on increased numbers of undifferentiated cells as they occupy a larger fraction of the space

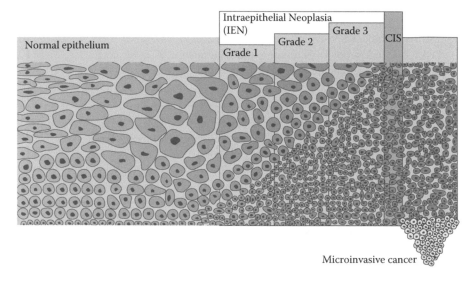

FIGURE 8.1 Schematic representation of normal tissue progressing through intraepithelial neoplasia (IEN) to invasive cancer. Note that in normal epithelium cells from the basement membrane mature as they proceed to the superficial epithelium. Grade 1 IEN is defined by cells remaining undifferentiated from the basement membrane to 1/3 the distance to the superficial epithelium. Grade 2 IEN is defined by cells remaining differentiated from 1/3 to 2/3 the distance from the basement membrane to the superficial epithelium. Grade 3 IEN is defined by undifferentiated cells being present in the area over 2/3 the distance from the basement membrane to the superficial epithelium. Carcinoma *in situ* (CIS) is defined as full thickness involvement from the basement membrane to the superficial epithelium. Counterintuitively microinvasive and invasive cancers invade the basement membrane and involve breaking into the stroma underlying the epithelium. Once the cancerous cells enter the stroma they can access both blood vessels and lymphatics. This process is related to metastasis. (Adapted from Goldstone, S. E. 2005. *The PRN Notebook* 10 (4): 11–16. With permission.)

between the basement membrane and the surface epithelium. The process becomes clinically critical when it seeks expansion in the opposite direction. Neoplastic cells invade the stroma (becoming invasive cancer) leading to their access to blood and lymphatic vessels that are necessary for metastasis to occur. This process, in most neoplasms, has been shown to occur over 20–30 years. Clearly intervening at any time while the lesion is precancerous is not only lifesaving but permits greatly less morbid treatments to be successful (Khanavkar 1998; Palcic 1991; BCCA 2005; Lane 2006).

Most cancers originate in sites that are directly accessible to screening and detection tools or are accessible by endoscopy. Data from the GLOBOCAN 2002 database (Ferlay 2004) shows that only 9.7% of cancers were recorded as nonepithelial in origin (leukemia, multiple myeloma, non-Hodgkin's lymphoma, or Hodgkin's lymphoma, brain and thyroid). The remaining 90% of cancers arise in organs with epithelia that can be palpated on physical exam and/or visualized using endoscopy, especially if one broadens the visualization devices to include laparoscopy, that can visualize and access the abdominal cavity (Heintzmann 2001). Thus, 90% of cancers should, one day, be preventable using optical technologies in the cancer screening and detection paradigm. Table 8.1 shows the endoscope-compatible technologies currently available for cancer screening and detection, their field of view, and resolution at the present time. In the screening paradigm, the characteristics of confocal endoscopy make it most suited to be a second-line imaging disease interrogation modality.

8.1.1 Cancer Screening and Detection

The well-established cancer screening paradigm is to observe an abnormality directly or indirectly through a cellular examination and if abnormal to proceed to histopathologic biopsy (Khanavkar 1998; Palcic 1991; BCCA 2005). Histopathologic confirmation of a condition likely to progress to cancer can lead to its local excision or treatment with a chemopreventive agent. Histopathologic confirmation of cancer leads to further testing to establish the extent of the disease, and then to treatment with surgery, chemotherapy, radiation therapy, or combinations thereof. Since most cancers have a 20–30 year development phase, most pass through a stage of being an identifiable at-risk lesion, which can often be treated with resection or ablation and/or chemoprevention, treatments far less morbid than those need to treat cancer. Chemoprevention agents are compounds that cause the high-risk preinvasive lesions to regress to a more normal tissue type and range from nutrients to antiviral vaccines to other medicinal compounds. Chemoprevention trials are conducted at research centers and are not yet the standard of care, but there is growing enthusiasm for mounting studies worldwide (BCCA 2005; Lane 2006; Rosin 2000, 2002; Zhang 1997; Heintzmann 2001).

8.1.2 Endoscopy in Screening and Diagnosis

There have been tremendous advances in white-light endoscopy with respect to both magnification (the enlargement) and resolution (smallest resolvable feature).

TABLE 8.1

Endoscope-Compatible Optical Technologies Undergoing Technology Assessment for Their Role in Cancer Screening and Detection

Field of View	Modality	Biology Visualized or Measured	Resolution	Cost	Regulatory Approval	Clinical Deployment
Wide	White-light endoscopy	Surface morphology	Detector dependent	$$	Yes	Standard of care (SOC)
	Fluorescence endoscopy	Collagen and elastin remodeling		$$$	Yes	Adjunctive. SOC in some sites
	Narrow-band imaging (NBI)	Neovascularization		$$	Yes	Adjunctive
Narrow	Endoscopic confocal microscopy	Subsurface tissue and cell morphology	1 μm	$$$	Yes	Trials in progress
	Optical coherence tomography (OCT)	Mucosal stratification, subsurface tissue morphology	5–15 μm	$$$	Ophthalmology, Intravascular imaging (Europe), skin (Europe)	Used clinically in ophthalmology. technical efficacy studies for other organ sites
	Fluorescence spectroscopy	Optical signature associated with epithelial and stromal biochemistry	Probe dependent	$	Yes	Evaluated in randomized clinical trials
	Elastic scattering (reflectance) spectroscopy	Optical signature associated with tissue and cellular morphology	Probe dependent	$	Yes	Technical efficacy studies

The naked eye is thought capable of detecting 125–165 μm lesions, while high-pixel count charge coupled devices incorporated into endoscopes have made it possible to discriminate 10–70 μm objects. High resolution endoscopes have increased from 100–200 thousand pixels per image to 850 thousand pixels per image. These same endoscopes also offer magnification ranging from 1.5 X to 115 X (Sidorenko 2004).

Clearly, these advances could increase the sensitivity of endoscopy by allowing clinicians to see microvasculature and surface changes in epithelium. As intraepithelial neoplasia develops, it secretes growth factors that increase the vasculature supporting the lesion. At the same time, the nuclear-to-cytoplasmic ratio increases within cells as less cytoplasm is produced and the amount of DNA within the nucleus can increase as well as its organizational packaging which affects nuclear size. The amounts of DNA can become so abnormal as to be labeled aneuploid, the best studied biomarker for cancer detection, diagnosis, and survival. The stroma underlying even preinvasive lesions can receive an influx of inflammatory cells and other changes in the underlying support structure such as collagen type, collagen cross-linking and elastin modifications, all preparing for further increases of lesion growth. In the cervix, whitelight endoscopy is called *colposcopy* and a green filter making vasculature more apparent has been routinely used for at least 40 years. Thus, endoscopy can be aided by the use of chromatic filters.

There are areas of the mucosa that have squamocolumnar junctions: cervical transformation zone, gastro-esophageal junction, gastric-small bowel junction, small bowel-colon junction, and fallopian tube-ovary junction. These are areas where squamous epithelium is in constant transition to columnar epithelium. Since these are areas of cell turnover, they are prone to increased genetic events (mutations) and in all organ systems predisposed to the development of metaplasia and often dysplasia.

8.1.3 ENDOSCOPIC OPTICAL TECHNOLOGIES IN SCREENING AND DIAGNOSIS

Endoscopic optical technologies are part of the "moving target" problem facing medicine in that technology is constantly changing. Ideally, all technologies should be subjected to rigorous technology assessment before diffusion and common use. However, as interesting technologies are described in technical journals or their use is described in even a few patients, these technologies get pushed to the forefront. Sometimes they become locally disseminated before they are properly assessed. Technology assessment involves measures of the biologic plausibility, technical efficacy, clinical effectiveness, patient acceptance, cost-effectiveness, provider acceptance, diffusion, and finally commercialization. For example, it would be ideal to know how devices work, in both screening and diagnostic populations, to perform randomized trials showing if the device improves diagnosis, if it is acceptable to the patient, and is cost-effective prior to implementation. However, science dissemination marches ahead of the logical technical assessment paradigm, and case reports and case series of new technologies force their addition to the arsenal of cancer screening and detection studies in cancer centers.

The opportunities for endoscopic optical technologies applied to the existing screening paradigm can be viewed as threefold: one strategy would be to use optical technology to augment the sensitivity of white-light endoscopy for a large field of view with or without the use of a contrast agent. A second strategy would be to use an optical technology housed in a probe, to increase the specificity of detection so that only areas abnormal both by endoscopy and by the probe were actually biopsied. Eventually, in the future, as endoscopic optical technologies are validated in adequately sized trials, they may be used in a third way, that is, as an "optical biopsy" mentioned in many reviews of these technologies.

There are confounders in the detection of cancer in low- and high-risk patients. Just as inflammation occurs in the stroma during the changes from precancers to cancer, there is inflammation of unknown etiology in most organ cites: examples exist such as esophagitis, gastritis, cervicitis, and colitis. In those organ sites, temporary inflammation is not regarded as a precursor to cancer, but in Barrett's disease of the esophagus, ulcerative colitis of the colon and Crohn's disease in the small bowel, there are increased risks of cancer associated with the increased chronic inflammation in these diseases.

Another confounder is neovascularization. Besides the influx of white cells, there is also neovascularization in inflammatory conditions, and the redness that is associated with the neovascularization visually, makes it very difficult to pick out areas to biopsy when examining these areas with endoscopy.

Two conditions are associated with both the influx of inflammation and the neovascularization; they are healing and benign inflammation. Inflammation and neovascularization also occur in healing after injury, ulceration, laceration, burns, and other intrusions into the epithelium. Healing and inflammation provide the greatest challenge to white-light endoscopy and perhaps the greatest opportunity for endoscopic optical technologies.

One difficulty is that in low-risk screening populations, the findings of cancer are rarer events, making it critically important that the procedures have high sensitivity. Here lies an opportunity for optical technologies to improve the sensitivity of white light scoping. Tests can be optimized for sensitivity. Since sensitivity and specificity are intrinsically linked, for any one technology one trades off sensitivity at the expense of specificity, meaning there may be more false-positives evaluated. Since optical technologies can be made to both view the large field of view as well as probe the exact location of an abnormality, it is possible to optimize a probe-based algorithm for specificity part during the tissue-contact part of the process. Using combined devices gives optical technologies a unique place in the cancer screening paradigm. This approach is being realized in the gastrointestinal tract using endoscopy combined with confocal microscopy.

Currently, it is well established that the diagnosis of cancer improves with the number of biopsies taken. The number of biopsies taken in the colon can number 50–60, especially if there are confounding inflammatory diseases like Barrett's esophagus, ulcerative colitis, and Crohn's disease. If that number could be reliably decreased by a probe with greater specificity, then fewer biopsies could be used to make the diagnosis. This strategy is being cautiously studied in the gastrointestinal tract. This alone would decrease the cost and morbidity of the cancer diagnostic process in this site.

8.2 CONTRAST AGENTS

In tandem with the increased resolution and magnification of endoscopes, there have been many studies using contrast agents to make the abnormal areas more visible when observing the wide angle view of the organ. There are three types of contrast agents: those applied topically, those used systemically, and those created with nanoparticles that bind to biomarkers of carcinogenesis in the cell. Most of these contrast agents also improve the contrast for confocal endoscopy. A list of vital contrast agents commonly employed during endoscopic confocal microscopy is shown in Table 8.2.

8.2.1 TOPICAL CONTRACT AGENTS

The vital dyes for which there is the most clinical data over the last 40–50 years, include, but are not limited, to methylene blue, acetic acid, Lugol's iodine, Toluidine blue, Cresyl violet, hypertonic saline, and indigo carmine (Sidorenko 2004). All of these agents are applied topically to the epithelium. Methylene blue stains actively absorbing cells and has been used in the oral cavity and gastrointestinal tract for many years. Acetic acid has been used in scoping the vulva, vagina, and cervix as well as in the esophagus. Lugol's iodine has an affinity to glycogen in non-keratinized epithelium and thus has established use in the esophagus and cervix. Toluidine blue has been used in the oral cavity and esophagus to identify areas of increased nuclear activity. Hypertonic saline has been used in the cervix as a contrast agent to enhance cytoplasmic size so that nuclear/cytoplasmic ratios could be calculated using the confocal microscope. It did not offer an advantage over acetic acid. Cresyl violet has been used as a topical antimicrobial agent and as a stain for abnormal areas in the gastrointestinal tract. Indigo carmine is a dye that can highlight irregularities in the mucosa and has been studied principally in the gastrointestinal tract. The Acridine dyes—Acriflavine, Proflavine, and Acridine Orange—stain cell nuclei by intercalation with nucleic acids (Acridine Orange is not generally used as a vital

TABLE 8.2
Vital Contrast Agents for Endoscopic Confocal Microscopy

Mode	Name	Affinity	Contrast Enhancement
Fluorescence	Fluorescein sodium	Albumin	Intercellular spaces (topical) and vasculature (IV injection)
	Acriflavine hydrochloride	DNA/RNA	Cell nuclei
	Proflavine	DNA/RNA	Cell nuclei
	Methylene blue	DNA	Cell nuclei
	Cresyl violet acetate	Cytoplasm	Negative enhancement of cell nuclei
Reflectance	Acetic acid	Nonspecific	Cell nuclei
	Hypertonic saline	Nonspecific	Cytoplasm
	Toluidine blue	DNA	Cell nuclei

dye because the intercalation is irreversible). Many of these contrast agents require that mucous covering the epithelium be washed away prior to use and/or require saline washes after application to optimize staining. Most unfortunately, there are few trials in which the investigators have used the same methodology for application, the same concentrations of contrast agents, or even in the same organ site, so that the quantification of the additional benefits of contrast agents is difficult to measure.

8.2.2 SYSTEMIC CONTRAST AGENTS

Systemic contrast agents are photosensitizers and can be given intravenously, orally, rectally, or as a gas. A photosensitizer is a chemical entity, which upon absorption of light induces a chemical or physical alteration of another chemical entity. Time must pass before the chemical transformation has taken place. Some photosensitizers are used both diagnostically and therapeutically such as in photodynamic therapy and for diagnosis of cancer (using fluorescence spectroscopy as part of endoscopy). Photodynamic therapy has been approved for the treatment of macular degenerations and also for several cancer indications (primarily in the bladder). Virtually all epithelial precancers have been studied using photosensitizers and fluorescence spectroscopy for diagnosis. The photosensitizers used are in most cases based on the porphyrin structure. The principal guiding their use for diagnosis or treatment is that they generally accumulate in precancerous and cancer epithelium to a higher extent than in the surrounding tissues and their fluorescing properties may be utilized for cancer detection. The photosensitizers may be chemically synthesized or induced endogenously by an intermediate in heme synthesis, 5-aminolevulinic acid (5-ALA) or 5-ALA esters. The most common therapeutic effect is based on the formation of reactive oxygen species upon activation of the photosensitizer by light (Verveer 1998; Hanley 1999; Berg 2005). These agents have been established and used over the last 20 years. Other agents that are also photosensitizers are mostly used for cancer staging and most of these have been used for over 20 years and include oral or rectal barium, oral or rectal Gastrogaffin, and then agents used intravenously: iodine-based, gadolinium, 18F-fluoro-deoxy-glucose, and Fluorescein. Fluorescein used intravenously adds good contrast to confocal endoscopy and is currently the subject of several gastrointestinal studies.

8.2.3 MOLECULAR CONTRAST AGENTS

The third type of contrast agent involves linking nanoparticles to receptors of known biomarkers of carcinogenesis. For example, we show three examples from the Richards-Kortum laboratory in Figure 8.2: that of anti-HPV 16 E7 antibodies, anti-EGFR gold bioconjugates, and antibodies to telomerases. In part A of Figure 8.2 we see binding to SiHa Cells and CaSki cells (known cervical cancer cell line derived from HPV infected cells) but principally not to C33-A (a non-HPV infected cell line) nor NHEK cells (normal human epithelial keratinocytes). On the right-hand part of Figure 8.2A are gold nanoparticles bound to the HPV antibodies that glow when illuminated with a laser pointer when the cell culture is under a simple phase-contrast microscopy. A second example from the Richards-Kortum laboratory is

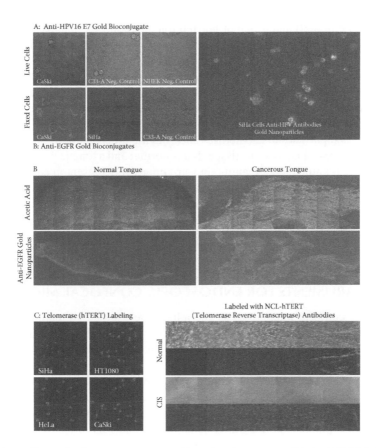

FIGURE 8.2 (See color insert) Imaging with molecular contrast. (A) Anti-HPV16 E7: Gold nanoparticles conjugated to anti-HPV16 E7 antibodies (Sokolov, K., J. Aaron, B. Hsu et al. 2003. *Technol Cancer Res Treat* 2 (6): 491–504). On the left, confocal reflectance images of live and fixed CaSki (cancerous), SiHa (cancerous), C33-A (normal) cervical cell lines, and NHEK (normal human epidermal keratinocyte) cell line cultured in the antibody stain. Minor staining is seen in the C33-A (HPV-negative) cell line which is known to be p16INK4A-positive. On the right, widefield reflectance images (using laser-pointer illumination) of SiHa cells cultured with the same antibody probe. (B) Anti-EGFR: Confocal reflectance images of normal (left) and cancerous (right) human tongue imaged *ex vivo* (Sokolov, K., J. Aaron, B. Hsu et al. 2003. *Technol Cancer Res Treat* 2 (6): 491–504). The tissue imaged in the top row was topically stained with 6% acetic acid while the tissue on the bottom was topically stained with gold nanoparticles conjugated to anti-EGFR. Normal tissue is EGFR-positive in the basal epithelium only whereas the cancerous specimen is EGFR-positive throughout entire specimen. (C) hTERT-telomerase reverse transcriptase: Detection of intra-nuclear markers by fluorescence following labeling with NCL-hTERT (Richards-Kortum, R., and M. Follen. 2006. Presentations to the National Institute of Biomedical Imaging and Bioengineering (NIBIB) by RRK and the the National Cancer Institute (NCI) by MF). On the left are cell lines imaged with a fluorescence microscopy. Images of SiHa (cancerous cervical cell line), HT1080 (cancerous fibrosarcoma cell line), HeLa (cancerous cervical cell line), and CaSki (cancerous cervical cell line) cells labeled NCL-hTERT antibody (green) are present. On the right are normal porcine cheek labeled with the same antibody; one sees minimal labeling in the normal tissue and full thickness labeling in the tongue with CIS.

shown in Figure 8.2B and is that of normal tongue tissue and tongue cancer stained both visualized with the confocal microscope and then imaged showing anti-EGFR (epidermal growth factor receptor) antibodies linked to gold nanoparticles. The confocal microscopy pictures show the differences in the organization of the tissue. The gold nanoparticles show that EGFR is found only in the basement membrane of normal tissue while it stains full thickness in the cancerous tongue sample. Finally, the third example from the Richards-Kortum laboratory is in Figure 8.2C, an example of binding nanoparticles to telomerases showing the green antibody NCL-hTERT in both cell cultures and in porcine tissue that is normal and a tongue with carcinoma *in situ*; notes that antitelomerase antibodies are intensively present in cancerous tissue (Goldstone 2005; van de Ven 2009; Sokolov 2003; Maitland 2008; Carlson 2007b). These three illustrations show us that contrast agents may become powerful tools for molecular imaging with confocal endoscopy once we can safely use them in humans. These examples are still in the laboratory and have not yet been approved or studied *in vivo* in humans.

8.3 INSTRUMENTS FOR ENDOSCOPIC CONFOCAL MICROSCOPY

Endoscopic confocal microscopy is the adjunctive use of a miniature endoscope-compatible confocal microscope with conventional (widefield) endoscopy. The confocal microscope derives its superior axial resolution by positioning an illumination and detection pinhole such that they are optically conjugate to a volume element of tissue under investigation. Thousands of these volume elements are imaged one by one in a raster pattern to build-up a complete image called an optical section. The principle of confocal fluorescence imaging using an optical fiber as the illumination and detection pinholes is illustrated schematically in Figure 8.3. The principle is similar for the case of confocal reflectance. The improvement in axial resolution is due to the ability of the confocal geometry to reject light that does not appear to come from the volume element being imaged. Confocal microscopy acquires 2-dimesional (2-D) optical sections from a 3-D object. More importantly, the focal plane for the probe can be stepped between sections perpendicular to the optical axis to build-up an entire 3-D image. This allows a confocal microscope to see both the epithelium and the stroma of an organ. Information from both adds to the increased sensitivity and specificity of these devices.

As shown in Figure 8.3, excitation light from the optical fiber (illumination pinhole) is projected into the tissue at the confocal plane. The light comes to a focus at the confocal plane but also illuminates tissue preceding the confocal plane and behind it. The excitation light induces fluorescence at the confocal plane (P) and in the tissue volume behind (P$^-$) and in front (P$^+$) of the confocal plane. The fluorescence emission is captured by the objective lens and projected back to the fiber. Fluorescence originating at the confocal plane is coupled into the fiber (detection pinhole) and transmitted to a detector while that originating behind (P$^-$) or in front (P$^-$) of the confocal plane is not coupled into the fiber and not detected. 2-D confocal imaging is achieved by either scanning the fiber in the image plane of the objective lens (distal scanning) or by using a fiber-optic image guide composed of multiple fibers and scanning proximal to the image guide.

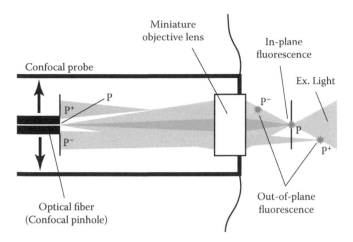

FIGURE 8.3 Principle of confocal fluorescence imaging in a fiber-optic probe. Excitation light from the optical fiber (illumination pinhole) is projected into the tissue at the confocal plane. The light comes to a focus at the confocal plane but also illuminates tissue before and after the plane. The excitation light induces fluorescence at the confocal plane (P) and in the tissue volume behind (P⁻) and in front (P⁺) of the confocal plane. The fluorescence emission is captured by the objective lens and projected back to the fiber. Fluorescence originating at the confocal plane is coupled into the fiber (detection pinhole) and transmitted to a detector while that originating behind (P⁻) or in front (P⁻) of the confocal plane is not coupled into the fiber and not detected. Two-dimensional confocal imaging is achieved by either scanning the fiber in the image plane of the objective lens (distal scanning) or by using a fiber-optic image guide composed of multiple fibers and scanning proximal to the image guide.

8.3.1 CLASSES OF INSTRUMENTATION

How the confocal pinhole is scanned is one of the primary differences between systems for endoscopic confocal microscopy. One can separate the systems into two main classes based upon the type of optical fiber. Broadly the two basic methodologies are: (1) Proximal scanning in which the scanning apparatus is coupled to a flexible fiber-optic image guide that conducts light to and from a miniature objective lens (Gmitro 1993; Sung 2002; Lane 2000; Knittel 2001); and (2) Distal scanning in which the light is conducted by a single optical fiber to and from the distal tip of the system, and the scanning of the illumination and detection pinhole (active area of the optical fiber) is accomplished by a miniaturized system at the distal end of the system (Tearney 1998; Dickensheets 1996; Hofmann 1999; Inoue 2000; Delaney 1994a, 1994b; Wang 2003). The various distal and proximal scanning approaches are illustrated schematically in Figure 8.4.

8.3.1.1 Proximal Scanning

There are two design forms for proximal scanning. Similar to the majority of benchtop confocal microscopes, these systems (Figure 8.4A) use one or more moving mirrors (galvanometers or resonance scanners) to scan the confocal pinholes in a raster fashion across the fiber-optic image guide (Sung 2002; Knittel 2001). A modification of this method uses a single mirror (Figure 8.4B) to scan confocal slits (instead

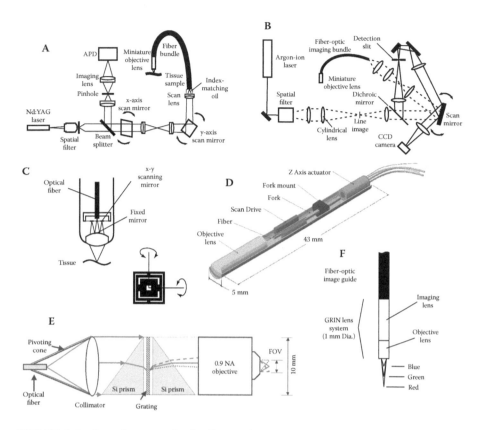

FIGURE 8.4 Scanning strategies for fiber-optic confocal endomicroscopy: Proximal scanning using a fiber-optic image guide (A-B) and distal scanning using a single optical fiber (C-F). (A) Spot-scanning; (B) slit-scanning; (C) microelectromechanical systems scanning; (D) resonance (tuning-fork) scanning; (E) lateral chromatic scanning; (F) Axial chromatic scanning. (Adapted from MacAulay, C., P. Lane, and R. Richards-Kortum. 2004. *Gastrointest Endosc Clin North America* 14 (3): 595–620. With permission.)

of pinholes) across the fiber-optic image guide (Gmitro 1993). This modification improves the speed of the imaging by parallelizing the illumination and detection pathways (a slit versus a pinhole) at the expense of some loss of the ability to remove out of focus light reducing the optical sectioning ability of the system as a whole. An alternate method is to use a MEMS (microelectromechanical system) device to behave as an array of independently controllable illumination pinholes that can very rapidly sequentially illuminate one or more fibers of the image guide fiber bundle. The fiber-optic image guide optically couples to a digital camera in which the individual pixels of the camera behave as the detection pinholes (Lane 2000).

8.3.1.2 Distal Scanning

There are at least three different design forms for *distal* scanning. The first makes use of a MEMS mirror or mirrors (Figure 8.4C) to scan the pinholes (as represented by the single optical fiber) across the tissue to be image (Dickensheets 1996; Wang

2003). Another method (Figure 8.4D) for distal scanning involves inducing the distal tip of the fiber to oscillate at its natural resonance frequency (Delaney 1994a, 1994b). Here the active area of the optical fiber acts as the pinhole and as projected through the objective lens raster scans the tissue plane to be imaged. A third method (Figure 8.4E) involves a combination of moving the optical fiber in one direction and using wavelength specific optical diffraction to create a line of pinholes, each of which conducts only a specific wavelength range of light to and from the targeted sample. The inventors of this technique at the Wellman Laboratories have called this technique spectrally encoded confocal microscopy (Tearney 1998). Similarly, axial chromatic scanning (Figure 8.4F) is possible by employing a distal objective lens with known axial chromatic aberration (Lane 2003).

8.3.2 CONTRAST LIMITS OF PROXIMAL SCANNING

Endoscopic confocal microscopes that employ proximal scanning must use a fiber-optic image guide to relay light to and from the distal objective lens. The resolution and contrast of this class of endomicroscope is generally limited by the optical fibers of the image guide. Resolution is limited by the number of optical fibers in the image guide. Currently available 1 mm diameter image guides from Sumitomo and Fujukura contain 30,000 fibers. This resolution is quite modest compared to state-of-the-art (widefield) endoscopes, which contain almost 1 million pixel elements. Contrast is also limited by the fiber-optic image guide. The mechanism of this limitation differs between reflectance and fluorescence modes of imaging.

In confocal fluorescence, the optical fibers exhibit background autofluorescence that reduce the signal-to-noise ratio (SNR) of the resulting images. A recent study characterized the autofluorescence of three commercial fiber-optic image guides (Udovich 2008). Using a custom monochromator designed for measuring tissue autofluorescence, excitation-emission matrices (EEMs) for each image guide were measured with excitation wavelengths ranging from 250 to 650 nm. The measurements from all three image guides reveled autofluorescence peaks at 400 nm emission due to excitation at 265 and 345 nm. The source of the autofluorescence was thought to originate from the Ge doping of the fibers in the image guide. Therefore, the best possible image contrast is achieved at wavelengths that avoid the image-guide autofluorescence peak at 400 nm. Autofluorescence still limits the SNR at other wavelengths.

In confocal reflectance, image-guide contrast is limited by Fresnel reflections at the input and output faces of the image guide and by Rayleigh scattering in the fibers. The dominant source is wavelength dependent: at shorter wavelengths the SNR is limited by Rayleigh scatter and at longer wavelengths the SNR is limited by Fresnel reflections. The intensity of the Fresnel reflections and the Rayleigh backscatter are of the same order of magnitude as the tissue backscatter under investigation (Lane 2009b). Consequently, this may limit the utility of confocal reflectance endoscopy without the use of contrast agents to increase the signal from the tissue. Using a refractive-index coupling fluid that accurately matches the refractive index of the fiber cores is critical to minimizing the Fresnel reflections from the end faces of the image guide (Lane 2009c). The Rayleigh scattering component can be accurately predicted using the fractional refractive-index difference and length of the fibers in the image guide.

8.3.3 CONTRAST MODES: REFLECTANCE AND FLUORESCENCE

Endoscopic confocal microscopes, like their bench-top counterparts, operate in one of two different contrast modes, namely reflectance or fluorescence. In reflectance mode, the same wavelength of light is used to illuminate the target as is detected from the target. This imaging mode makes use of differences in tissue component backscattering characteristics that either are native to the tissue or induced by a contrast agent such as acetic acid to modify the optical properties of the cellular components to be imaged. Additional vital stains for confocal reflectance are listed in Table 8.2. As with all devices for endoscopic confocal imaging, the goal is to differentiate between the nuclear component of cells and the cytoplasmic component so as to be able to visualize the cell nuclei.

Confocal fluorescence imaging requires that an optically active fluorescence dye is delivered to the tissue. The vital fluorescent dye most commonly reported in the literature is intravenous fluorescein sodium at a concentration of between 1 and 10% (topical application of fluorescein sodium is also possible). Fluorescein highlights the vasculature, lamina propria, and intracellular spaces, allowing the visualization of vessel pattern and cellular architecture. Acriflavine hydrochloride (0.05%) has been used as a topical stain and has the benefit of staining nuclei which are not stained by fluorescein sodium. Additional vital stains for confocal fluorescence are listed in Table 8.2. Although both of these dyes have been used clinically for some time (fluorescein sodium is FDA approved for angiography and acriflavine hydrochloride is used as a topical antiseptic), the safety of acriflavine hydrochloride is sometimes called into question because of its ability to (reversibly) intercalate DNA. Confocal (auto) fluorescence is also possible using the endogenous (naturally occurring) fluorophores present in tissue. This enables one to image the connective tissues such as collagen cross-links and elastin that make up the basement membrane and submucosa of most epithelial tissues.

The minimum imaging requirements to resolve individual cells *in vivo* are a lateral resolution approximately 3 µm or less (preferably 1 µm) in the transverse directions and an optical sectioning ability (axial resolution) on the order of 10 µm or better. For resolving larger targets such as microvilli in the gastrointestinal tract or tissue microvascularization, substantially less resolving capability is required. Tightly coupled to this specification is the achievable field of view. Generally, larger is better. However, resolution, field of view and imaging time are all very tightly coupled, and it is difficult to improve upon one without having to sacrifice on the others. For cellular-level imaging, a field of view approximately 200 × 200 µm is likely sufficient to include enough structural and organization information for the visual assessment of the imaged tissue. The ability to perform optical sectioning *in vivo* (i.e., resolve information from a specific level within the tissue implies the need to be able to control from which specific level within the tissue the information is obtained). Among the systems currently under development three primary methods have emerged: moving optical components within the distal lens system to change the optical path length to the focal plane within the tissue, moving the tissue relative to the distal optical system, and spectrally selecting focal depth using a lens system intentionally designed with axial chromatic aberration such that the wavelength of the light determines the depth from which the image comes (Figure 8.4F).

8.3.4 RESOLUTION

The confocal optical sectioning ability of all these systems depends upon the NA (numerical aperture) of their distal objective lens and the size of the confocal pinhole. However, high NA optics tends to be larger and the diameter of the distal end of these confocal microendoscopes is tightly limited with respect to being able to reach the organs and sites they are targeted for. For most of these systems, it is the size of the distal optics that limits how small the probes may be constructed. The probes for currently available commercial systems range in diameter from 0.3 to 12.8 mm (see Table 8.3). Given the size of the objects being resolved relative to the size of the probe and organ in which they reside, it is surprising that, for most of these systems, motion artifacts are not an insurmountable problem. To work properly, these systems need to be locationally stable with respect to the tissue they are imaging to within a few microns over a substantial fraction of a second. The frame rates for commercial systems range from 0.8 to 12 frames per second (see Table 8.3). As a result, the probe tissue geometry needs to stay stable to within a few microns for over a second. Seemingly difficult, it turns out that the probes flexibility and, once in contact with the tissue, they tend to drag the surface layer of the tissue in contact with the probe along with any motion of the probe. Since most of these systems can only image from the surface to about 250 μm into the tissue, it is only the macroscopic surface layer that is being imaged.

8.4 REVIEW OF THE CLINICAL LITERATURE

A review of the clinical literature was conducted using Medline and PubMed. The search terms "*in vivo* confocal" and "confocal endoscopy" were crossed with "reviews," "methodology," "instrumentation," "animal models," "clinical trials," "bladder," "oral cavity," "cervix," "lung," "gastrointestinal tract," "endometrium," and "ovary." A total of 120 articles were found and their references reviewed.

A broader search of company websites, Optiscan (Notting Hill, Australia, www.optiscan.com) and Mauna Kea Technologies (Paris, France; www.maunakeatech.com), led to the review of an additional 260 abstracts, journal articles, and book chapters that did not come up in the original search. The search terms used in these articles were "*in vivo* laser scanning microscopy," "confocal laser endomicroscopy," "chromoscopic endomicrosopy," "high-magnification chromoscopic colonoscopy," and "confocal mini-probe." This lack of equivalent search terms is a characteristic of emerging technologies. We have eliminated the duplicate references, the preclinical references, book chapters, abstracts, and research presentations and kept the peer-reviewed English literature in this review.

TABLE 8.3

Commercial Systems for *In Vivo* Confocal Microscopy

Vendor	System		Regulatory Approval	Probe Diameter [mm]	Contrast Mode[1]	λ [nm]	Resolution [μm]		FOV [um]	Working Distance [μm]	Frame Rate[2] [Hz]
							Lateral	Axial			
Pentax (OptiScan)	EC-3870CIK (ISC-1000)		FDA, CE	12.8	F	488	0.7	7	475 (sq)	0–250	0.8
OptiScan	Five 1			6.3	F	488	0.7	7	475 (sq)	0–250	0.8
Mauna Kea	Cellvizio	S Series		0.3–1.5	F	488 or 660	3.5	15	300–600	0	12
	LAB	HD Series		1.8			2.5	20	240	30, 50, 80	
		MiniO		4.2			1.8	5	240	30	
	Cellvizio	Cholangioflex	FDA, CE	1	F	488	3.5	30	320	55	12
	GI	Gastroflex[UHD]		2.7			1	10	240	60	
		Gastroflex		2.6			3.5	60	600	100	
	Cellvizio Lung	AlveoFlex	FDA, CE	1.4	F	488	3.5	50	600	25	12
Lucid	VivaScope	1500 3000	FDA	Arm Hand-held	R	830	<2	<5	500 (sq)	0–400	9
	VivaCell	2500	N/A	Desktop	R	830		<5	500 (sq)	0–350	9

Note: (1) Contrast mode is either confocal fluorescence (F) or confocal reflectance (R). (2) Frame rate based on acquisition of 1024x1024 pixel images. FDA = United States Food and Drug Administration (FDA) Class II regulatory clearance for microscopic visualization. CE = Conformity with European Union safety and health requirements. FOV = Field of view.

The clinical literature review is summarized in Table 8.4. The articles are classified as reviews (if the term "review" was used in the title), pre-clinical (cell culture, *ex vivo* tissue, and animal studies) and clinical (*in vivo* human studies). Peer-reviewed articles were further classified by organ site: bladder, oral cavity, cervix, lung, and gastrointestinal tract. The table itself gives a picture of the clinical enthusiasm for the device. Two papers looked at *ex vivo* bladder tumors and there is one *in vivo* study. Confocal endoscopy has been studied by some investigators using acetic acid as a contrast agent and others not using any contrast agent, all for the diagnosis of neoplasia. Similarly, in the cervix, all of the studies focus on the use of confocal to diagnose neoplasia and many used acetic acid as a contrast agent. There is growing enthusiasm for confocal endoscopy in the lung, again as a second modality following fluorescence and reflectance bronchoscopy. There is clearly currently an overwhelming use of confocal endoscopy in the gastrointestinal tract. In the gastrointestinal tract, there are studies combing confocal endoscopy with other optical technologies and there are extensive studies of contrast agents.

Figure 8.5 demonstrates images taken from the bronchial tree of the lung. On the left panel are confocal images of a cancer stained with Cresyl violet and the corresponding H&E image. On the right are images from Thiberville showing the differences between smokers and nonsmokers evident with confocal endoscopy. The detail is exquisite and is believed due to changes in elastin. Figure 8.6 shows images of the human cervix showing work both *ex vivo* and *in vivo* in areas that are normal, dysplastic, and showing CIS. It's believed that confocal cervical endoscopy could eliminate the confounders see in colposcopy and in fluorescence and reflectance spectroscopy. In Figure 8.7, there are images of the oral cavity both *ex vivo* and *in vivo*. Here are examples of normal tissue, CIS, and invasive cancer seen with confocal imaging. Similar to the cervix, this epithelium can be easily accessed with confocal probes that could target areas for biopsy or determine surgical margins. Figure 8.8 shows confocal endomicroscopy used to identify normal mucosa in the stomach, intestinal metaplasia, Barrett's esophagitis, and a tubular adenoma in the colon. Figure 8.9 contrasts normal, high-grade intra-epithelial neoplasia (IEN), and cancer imaged in the colon by confocal endomicroscopy using intravenous fluorescein as a contrast agent. The corresponding tissue morphology is indicated in the accompanying Hematoxylin and Eosin (H&E) stained sections.

Commercial systems for *in vivo* confocal microscopy are presented in Table 8.3 along with their key specifications. Pentax (Tokyo, Japan; www.pentaxmedical.com) sells a colonoscope with an integrated confocal microscope (EC-3870CIK). The confocal processor of the Pentax colonoscope (ISC-1000) is licensed from Optiscan Pty. Ltd. (Notting Hill, Australia; www.optiscan.com), which also sells a different *in vivo* confocal microscope based on the same technology (Five 1) but which is not endoscope-compatible. Mauna Kea Technologies (Paris, France; www.maunakea-tech.com) sells miniature confocal probes (Cellvizio) that are compatible with the instrument channel of most colonoscopes, gastroscopes, duodenoscopes, and bronchoscopes. Lucid Technologies (Rochester, New York; www.lucid-tech.com) also sells several instruments for *in vivo* confocal imaging, however, the Lucid systems are not endoscope-compatible.

A search of ClinicalTrials.gov, the U.S. National Institute's of Health clinical trial database, revealed 25 open clinical trials using *in vivo* confocal endoscopy. Studies

TABLE 8.4

Literature Review: Clinical Application of Endoscopic Confocal Microscopy by Study Type and Organ Site

Organ Site	Bladder	Oral Cavity	Cervix	Lung	Gi Tract
Reviews			Bazant-Hegemark 2008; Drezek 2003; Tan 2007		Kiesslich 2006e; Atkinson 2007; Kiesslich 2006c; Van den Broek 2007; Kiesslich 2007g, 2009; Hurlstone 2006; Hurlstone 2007a; Evans 2005; MacAulay 2004; Gossner 2008; Dunbar 2008; Kiesslich 2005b; Reddymasu 2008; Kiesslich 2005d; Nguyen 2008; Kiesslich 2005c; Dacosta 2002
Preclinical (*ex vivo*, animal)	D'Hallewin 2005; Sonn 2009b	Carlson 2007a; Chapple 2000; Clark 2002, 2003; Paris 2009; Pavlova 2008)	Clark 2002; Collier 2002, 2000; Drezek 2000		Goetz 2009b; Delaney 1994b; Kopacova 2009; Goetz 2007a; Inoue 2000; Papworth 1998; Goetz 2009a; Kiesslich 2007b; Goetz 2007b

Clinical (*in vivo*)	(Sonn 2009a)	Haxel 2009; Just 2009; Maitland 2008; Thong 2007; Zheng 2004	Carlson 2005; Sung 2003; Tan 2009	Lane 2009a, 2007; Thiberville 2009b, 2009a	Meining 2007b; Dzierzanowska-Fangrat 2006; Kiesslich 2006b; Liu 2008; Pech 2008; Gheorghe 2008; Liu 2009; Kiesslich 2007e; Leung 2009; Bani-Hani 2008; Hurlstone 2008b; Cotruta 2009; Becker 2008b; Goetz 2008a; Zambelli 2008a; Meining 2007d; vonDelius 2007; Hurlstone 2008c; Bojarski 2009; Kiesslich 2008a; Hsiung 2008; Goetz 2006; Kiesslich 2007d; Guo 2008; Polglase 2006; Goetz 2008b; Kiesslich 2006d; Sanduleanu 2009; Kiesslich 2006f; Zambelli 2008b; Hurlstone 2008a; Dunbar 2009; Odagi 2007; Polglase 2005; Curvers 2008; Wang 2005; Kiesslich 2004, 2007a; Zhang 2007, 2008; Kiesslich 2007c, 2008b; Becker 2008a; Meining 2008b; Yeoh 2007; Buchner 2008; Kiesslich 2006a; Nathanson 2004; Becker 2007; Meining 2008a; Trovato 2007; Meining 2009; Hurlstone 2007c; Meining 2007c; Hoffman 2006; Wang 2007; Hurlstone 2007b; Wallace 2009; Pohl 2008; Kiesslich 2007f; Leong 2008; Kiesslich 2005a; Venkatesh 2009b; Monkemuller 2009; Morgner 2007; Venkatesh 2009a; Meining 2007a; Miehlke 2007: Canto 2006; Watanabe 2008
Article count	3	11	10	4	97

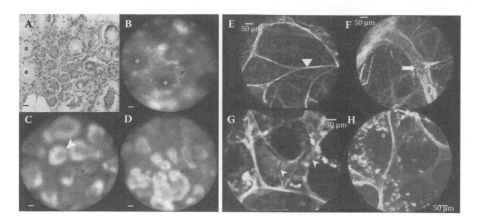

FIGURE 8.5 Confocal fluorescence images acquired from the human bronchial tree in the lung. Left panel (modified from Lane, P. M., S. Lam, A. McWilliams, J. C. Leriche, M. W. Anderson, and C. E. Macaulay. 2009a. *J Biomed Opt* 14 (2): 024008. With permission) illustrates cancer as stained by Cresyl violet: (A) H&E stained section indicating glandular carcinoma; (B–D) Corresponding confocal endoscopy images. The larger glandular structures (stars) correspond well to those in the H&E section. Smaller glandular structures with a larger cellular component are also indicated (arrows). Right panel (modified from Thiberville, L., M. Salaun, S. Lachkar et al. 2009a. *Proc Am Thorac Soc* 6 (5): 444–449. With permission) illustrates autofluorescence imaging (488 nm excitation) for a nonsmoker (E–F) and a smoker (G-H): (E) Elastin framework of an alveolar duct (arrowhead); (F) Extra-alveolar microvessel (arrow); (G) Alveolar walls (arrowhead); (H) Alveolar macrophages (asterisk).

included the oral cavity, cervix, biliary duct, eye, colon, lung, skin, prostate, and esophagus. Most of the trials were focused on the evaluation of colorectal cancer, evaluation of ulcerative colitis and dysplasia, the diagnosis of oral cancer, diagnosis of cervical cancer, evaluation of the etiology of biliary duct strictures, evaluation of interstitial lung disease diagnoses, margin delineation in robotic-assisted prostatic cancer resection, evaluation of Barret's esophagus, evaluation of a treatment for pseudofolliculitis barbae, evaluation of the effects of filters on macular degeneration, and the evaluation of lacrimal gland secretion in Sjogren's Disease (Partridge 1998).

Mauna Kea has registered trials with the National Cancer Institute recruiting patients for cholangiopancreatography for the diagnosis of pancreatic and bile duct cancers. Further searches of funded "clinical trials and imaging" from the National Institute of Biomedical Imaging Branch showed 41 funded grants and from the National Cancer Institute; the search yielded 304 funded studies. There was little but not complete overlap in these lists so that one can assume over 300 grants have been funded in the past or present for "imaging and clinical trials" in 2009 (Califano 1996).

8.5   QUANTITATIVE IMAGE ANALYSIS

Confocal endomicroscopy systems can produce images of cell nuclei from several hundreds of cells. The skilled observer with a familiarity of the histopathological changes associated with the diseases states being examined can interpret this

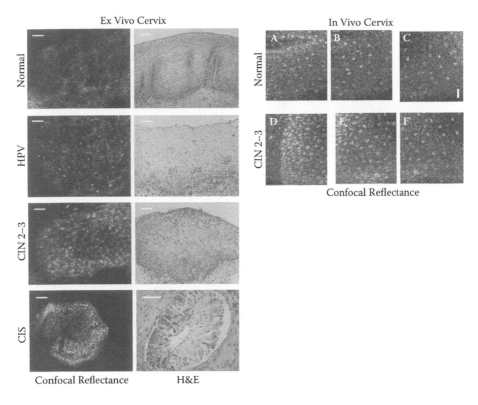

FIGURE 8.6 On the left are *ex vivo* images of confocal reflectance images of the human cervix using the contrast agent acetic acid and associated Hematoxylin and Eosin (H&E) sections (modified from Collier, T., A. Lacy, R. Richards-Kortum, A. Malpica, and M. Follen. 2002. *Academic Radiol* 9 (5): 504–512. With permission). On the right are *in vivo* confocal reflectance images of the cervix using the contrast agent acetic acid images for which the histopathology has been confirmed. (Modified from Sung, K. B., C. Liang, M. Descour et al. 2002. *J. Microscopy-Oxford* 207:137–145. With permission.)

population of cells. Figure 8.10 shows a confocal image from the lung and a graph of quantitative histopathological lung tissue score showing that normal tissue and cancer can be easily separated. It is now possible to conduct multicenter studies using industrial grade devices of particular precancer and cancers with sufficient power in the study design, central histopathological review, quantitative histopathological study, biomarker study, contrast agent study, and data collection of images. Studies should be powered to produce sufficient data that datasets can be divided into training, validation, and test sets. These studies require careful design, administration, conduct, and evaluation. Further, there are groups developing automated quantitative analysis tools to examine the spatial arrangement of cells in tissue to assist in the diagnostic evaluation of the images produced by confocal endomicroscopy systems. These tools are similar to those developed and used in automated image analysis histology slides. Despite the 3-D nature of the technology, some groups have shown that 2-D views provide adequate information so that diagnostic algorithms can be derived.

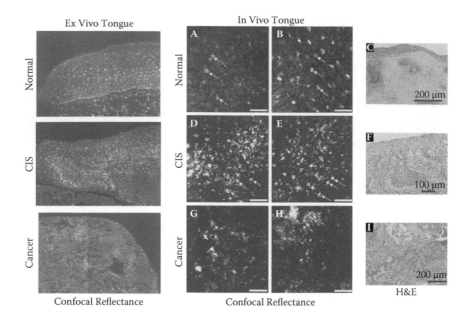

FIGURE 8.7 Confocal reflectance images of *ex vivo* and *in vivo* (right, modified from Maitland, K. C., A. M. Gillenwater, M. D. Williams, A. K. El-Naggar, M. R. Descour, and R. R. Richards-Kortum. 2008. *Oral Oncol* 44 (11): 1059–1066. With permission) human tongue. Acetic acid was used as a contrast agent and tissue grade (normal, CIS and cancer) was confirmed by histopathology. H&E sections are shown for the *in vivo* cases only.

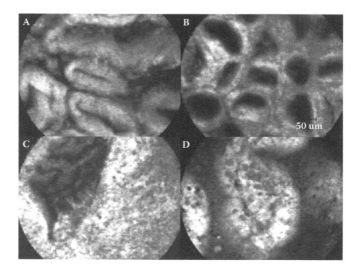

FIGURE 8.8 Confocal fluorescence endomicroscopy images from video sequences of the gastrointestinal tract, showing hyperplastic (A) and normal cardia mucosa (B), specialized intestinal metaplasia (BE) adjacent to squamous epithelium (C), and a tubular adenoma in the colon (D). Field of view is 500 x 600 µm. (Modified from Meining, A., D. Saur, M. Bajbouj et al. 2007c. *Clin Gastroenterol Hepatol* 5 (11): 1261–1267. With permission.)

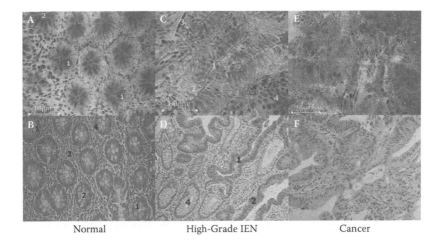

Normal   High-Grade IEN   Cancer

FIGURE 8.9 Confocal endomicroscopy (top) and corresponding H&E stained sections (bottom) of normal (left), neoplastic (middle), and cancerous (right) colorectal mucosa. (A) Confocal section of normal mucosa showing round-shaped regular colonic crypts with black mucin visible within goblet cells (1 = goblet cells, 2 = crypt lumen, 3 = stroma, 4 = nuclei) and (B) corresponding histology; (C) Confocal section of high-grade intra-epithelial neoplasia (IEN) showing tubular-shaped crypts with a reduced number of goblet cells and loss of cellular junctions (1 = branched crypt structure in the area of IEN, 2 = reduced number of goblet cells, 3 = loss of cellular junctions, 4 = normal-shaped crypts) and (D) corresponding histology. (E) Confocal section of cancer showing irregular cell architecture with total loss of goblet cells and (F) corresponding histology. (Modified from Kiesslich, R., J. Burg, M. Vieth et al. 2004. *Gastroenterology* 127 (3): 706–713. With permission.)

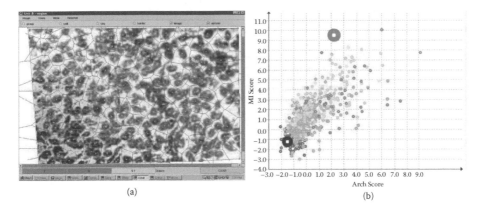

(a)   (b)

FIGURE 8.10 Quantitative image analyses. (A) Confocal reflectance image automatically processed using quantitative histopathologic architectural software. (B) Scatter plot of ArchScore (function based on architectural features) and MIScore (based on morphologic features) for a variety of tissues sections (lung, colon) in different stages of neoplastic progression (normal, hyperplasia, metaplasia, mild, moderate, severe dysplasia, CIS, and invasive cancer). Large black donut represents score from normal colon; large gray donut represents score from invasive colon cancer.

REFERENCES

Atkinson, R. J., A. J. Shorthouse, and D. P. Hurlstone. 2007. Novel colorectal endoscopic *in vivo* imaging and resection practice: A short practice guide for interventional endoscopists. *Tech Coloproctol* 11 (1): 7–16.

Bani-Hani, K. E., and B. K. Bani-Hani. 2008. Columnar lined (Barrett's) esophagus: Future perspectives. *J Gastroenterol Hepatol* 23 (2): 178–191.

Bazant-Hegemark, F., K. Edey, G. R. Swingler, M. D. Read, and N. Stone. 2008. Review: Optical micrometer resolution scanning for non-invasive grading of precancer in the human uterine cervix. *Technol Cancer Res Treat* 7 (6): 483–496.

BCCA. 2005. The Cervical Cancer Screening Program 2005 Annual Report. Vancouver: BC Cancer Agency.

Becker, V., T. Vercauteren, C. H. von Weyhern, C. Prinz, R. M. Schmid, and A. Meining. 2007. High-resolution miniprobe-based confocal microscopy in combination with video mosaicing (with video). *Gastrointest Endosc* 66 (5): 1001–1007.

Becker, V., M. Vieth, M. Bajbouj, R. M. Schmid, and A. Meining. 2008a. Confocal laser scanning fluorescence microscopy for *in vivo* determination of microvessel density in Barrett's esophagus. *Endoscopy* 40 (11): 888–891.

Becker, V., S. von Delius, M. Bajbcouj, A. Karagianni, R. M. Sclunid, and A. Meining. 2008b. Intravenous application of fluorescein for confocal laser scanning microscopy: Evaluation of contrast dynamics and image quality with increasing injection-to-imaging time. *Gastrointest Endosc* 68 (2): 319–323.

Berg, K., P. K. Selbo, A. Weyergang et al. 2005. Porphyrin-related photosensitizers for cancer imaging and therapeutic applications. *J Microsc* 218 (Pt 2): 133–147.

Bojarski, C., U. Gunther, K. Rieger et al. 2009. *In vivo* diagnosis of acute intestinal graft-versus-host disease by confocal endomicroscopy. *Endoscopy* 41 (5): 433–438.

Buchner, A. M., and M. B. Wallace. 2008. Future expectations in digestive endoscopy: Competition with other novel imaging techniques. *Best Pract Res Clin Gastroenterol* 22 (5): 971–987.

Califano, J., P. vanderRiet, W. Westra et al. 1996. Genetic progression model for head and neck cancer: Implications for field cancerization. *Cancer Res* 56 (11): 2488–2492.

Canto, M. 2006. Diagnosis of Barrett's esophagus and esophageal neoplasia: East meets west. *Dig Endosc* 18 (s1): S36–S40.

Carlson, A. L., A. M. Gillenwater, M. D. Williams, A. K. El-Naggar, and R. R. Richards-Kortum. 2007a. Confocal microscopy and molecular-specific optical contrast agents for the detection of oral neoplasia. *Technol Cancer Res Treatment* 6 (5): 361–374.

Carlson, A. L., A. M. Gillenwater, M. D. Williams, A. K. El-Naggar, and R. R. Richards-Kortum. 2007b. Confocal microscopy and molecular-specific optical contrast agents for the detection of oral neoplasia. *Technol Cancer Res Treat* 6 (5): 361–374.

Carlson, K., I. Pavlova, T. Collier, M. Descour, M. Follen, and R. Richards-Kortum. 2005. Confocal microscopy: Imaging cervical precancerous lesions. *Gynecol Oncol* 99 (3 Suppl. 1): S84–S88.

Chapple, C. C., R. K. Kumar, and N. Hunter. 2000. Vascular remodelling in chronic inflammatory periodontal disease. *J Oral Pathol Med* 29 (10): 500–506.

Clark, A., T. Collier, A. Lacy et al. 2002. Detection of dysplasia with near real time confocal microscopy. *Biomed Sci Instrum* 38: 393–398.

Clark, A. L., A. M. Gillenwater, T. G. Collier, R. Alizadeh-Naderi, A. K. El-Naggar, and R. R. Richards-Kortum. 2003. Confocal microscopy for real-time detection of oral cavity neoplasia. *Clin Cancer Res* 9 (13): 4714–4721.

Collier, T., A. Lacy, R. Richards-Kortum, A. Malpica, and M. Follen. 2002. Near real-time confocal microscopy of amelanotic tissue: Detection of dysplasia in *ex vivo* cervical tissue. *Academic Radiol* 9 (5): 504–512.

Collier, T., P. Shen, B. de Pradier et al. 2000. Near real time confocal microscopy of amelanotic tissue: Dynamics of aceto-whitening enable nuclear segmentation. *Opt Express* 6 (2): 40–48.

Cotruta, B., C. Gheorghe, and I. Bancila. 2009. Magnifying endoscopy with narrow-band imaging or confocal laser endomicroscopy for *in vivo* rapid diagnostic of Barrett's esophagus. *J Gastrointestin Liver Dis* 18 (2): 258–259.

Curvers, W. L., R. Kiesslich, and J. J. G. H. M. Bergman. 2008. Novel imaging modalities in the detection of oesophageal neoplasia. *Best Pract Res Clin Gastroenterol* 22 (4): 687–720.

D'Hallewin, M. A., S. El Khatib, A. Leroux, L. Bezdetnaya, and F. Guillemin. 2005. Endoscopic confocal fluorescence microscopy of normal and tumor bearing rat bladder. *J Urol* 174 (2): 736–740.

Dacosta, R. S., B. C. Wilson, and N. E. Marcon. 2002. New optical technologies for earlier endoscopic diagnosis of premalignant gastrointestinal lesions. *J Gastroenterol Hepatol* 17: S85–S104.

Delaney, P. M., M. R. Harris, and R. G. King. 1994a. Fiberoptic laser-scanning confocal microscope suitable for fluorescence imaging. *Appl Opt* 33 (4): 573–577.

Delaney, P. M., R. G. King, J. R. Lambert, and M. R. Harris. 1994b. Fibre optic confocal imaging (FOCI) for subsurface microscopy of the colon *in vivo*. *J Anat* 184 (Pt 1): 157–160.

Dickensheets, D. L., and G. S. Kino. 1996. Micromachined scanning confocal optical microscope. *Opt Lett* 21 (10): 764–766.

Drezek, R. A., T. Collier, C. K. Brookner et al. 2000. Laser scanning confocal microscopy of cervical tissue before and after application of acetic acid. *Am J Obstet Gynecol* 182 (5): 1135–1139.

Drezek, R. A., R. Richards-Kortum, M. A. Brewer et al. 2003. Optical imaging of the cervix. *Cancer* 98 (9): 2015–2027.

Dunbar, K. B., P. Okolo, 3rd, E. Montgomery, and M. I. Canto. 2009. Confocal laser endomicroscopy in Barrett's esophagus and endoscopically inapparent Barrett's neoplasia: A prospective, randomized, double-blind, controlled, crossover trial. *Gastrointest Endosc* 70 (4): 645–654.

Dunbar, K., and M. Canto. 2008. Confocal endomicroscopy. *Curr Opin Gastroenterol* 24 (5): 631–637.

Dzierzanowska-Fangrat, K., P. Lehours, F. Megraud, and D. Dzierzanowska. 2006. Diagnosis of Helicobacter pylori infection. *Helicobacter* 11: 6–13.

Evans, J. A., and N. S. Nishioka. 2005. Endoscopic confocal microscopy. *Curr Opin Gastroenterol* 21 (5): 578–584.

Ferlay, J., F. Bray, P. Pisani, and D. M. Parkin. 2004. *GLOBOCAN 2002: Cancer Incidence, Mortality and Prevalence Worldwide* (IARC CancerBase No. 5. version 2.0). IARC Press. Available from http: //www-dep.iarc.fr/.

Gheorghe, C., R. Iacob, G. Becheanu, and M. Dumbrava. 2008. Confocal endomicroscopy for *in vivo* microscopic analysis of upper gastrointestinal tract premalignant and malignant lesions. *J Gastrointest Liver Dis* 17 (1): 95–100.

Gmitro, A. F., and D. Aziz. 1993. Confocal microscopy through a fiberoptic imaging bundle. *Opt Lett* 18 (8): 565–567.

Goetz, M., C. Fottner, E. Schirrmacher et al. 2007a. In-vivo confocal real-time mini-microscopy in animal models of human inflammatory and neoplastic diseases. *Endoscopy* 39 (4): 350–356.

Goetz, M., A. Hoffman, P. R. Galle, M. F. Neurath, and R. Kiesslich. 2006. Confocal laser endoscopy: New approach to the early diagnosis of tumors of the esophagus and stomach. *Future Oncol* 2 (4): 469–476.

Goetz, M., and R. Kiesslich. 2008a. Confocal endomicroscopy: *In vivo* diagnosis of neoplastic lesions of the gastrointestinal tract. *Anticancer Res* 28 (1B): 353–360.

Goetz, M., R. Kiesslich, H. P. Dienes et al. 2008b. *In vivo* confocal laser endomicroscopy of the human liver: A novel method for assessing liver microarchitecture in real time. *Endoscopy* 40 (7): 554–562.

Goetz, M., B. Memadathil, S. Biesterfeld et al. 2007b. *In vivo* subsurface morphological and functional cellular and subcellular imaging of the gastrointestinal tract with confocal mini-microscopy. *World J Gastroenterol* 13 (15): 2160–2165.

Goetz, M., T. Toermer, M. Vieth et al. 2009a. Simultaneous confocal laser endomicroscopy and chromoendoscopy with topical cresyl violet. *Gastrointest Endosc* 70 (5): 959–968.

Goetz, M., A. Ziebart, S. Foersch et al. 2009b. *In vivo* molecular imaging of colorectal cancer with confocal endomicroscopy of epidermal growth factor receptor. *Gastroenterology* 138 (2): 435–446.

Goldstone, S. E. 2005. Diagnosis and treatment of HPV-related squamous intraepithelial neoplasia in men who have sex with men. *The PRN Notebook* 10 (4): 11–16.

Gossner, L. 2008. Potential contribution of novel imaging modalities in non-erosive reflux disease. *Best Pract Res Clin Gastroenterol* 22 (4): 617–624.

Guo, Y. T., Y. Q. Li, T. Yu et al. 2008. Diagnosis of gastric intestinal metaplasia with confocal laser endomicroscopy *in vivo*: A prospective study. *Endoscopy* 40 (7): 547–553.

Hanley, Q. S., P. J. Verveer, M. J. Gemkow, D. Arndt-Jovin, and T. M. Jovin. 1999. An optical sectioning programmable array microscope implemented with a digital micromirror device. *J Microscopy-Oxford* 196: 317–331.

Haxel, B. R., M. Goetz, R. Kiesslich, and J. Gosepath. 2009. Confocal endomicroscopy: A novel application for imaging of oral and oropharyngeal mucosa in human. *Eur Arch Otorhinolaryngol.* 267 (3): 443–448.

Heintzmann, R., Q. S. Hanley, D. Arndt-Jovin, and T. M. Jovin. 2001. A dual path programmable array microscope (PAM): Simultaneous acquisition of conjugate and non-conjugate images. *J Microscopy-Oxford* 204: 119–135.

Hoffman, A., M. Goetz, M. Vieth, P. R. Galle, M. F. Neurath, and R. Kiesslich. 2006. Confocal laser endomicroscopy: Technical status and current indications. *Endoscopy* 38 (12): 1275–1283.

Hofmann, U., S. Muehlmann, M. Witt, K. Doerschel, R. Schuetz, and B. Wagner. 1999. Electrostatically driven micromirrors for a miniaturized confocal laser scanning microscope. Proc. SPIE vol. 3878, Miniaturized Systems with Micro-Optics and MEMS, 29.

Hsiung, P. L., J. Hardy, S. Friedland et al. 2008. Detection of colonic dysplasia *in vivo* using a targeted heptapeptide and confocal microendoscopy. *Nat Med* 14 (4): 454–458.

Hurlstone, D. P., W. Baraza, S. Brown, M. Thomson, N. Tiffin, and S. S. Cross. 2008a. *In vivo* real-time confocal laser scanning endomicroscopic colonoscopy for the detection and characterization of colorectal neoplasia. *Br J Surg* 95 (5): 636–645.

Hurlstone, D. P., and S. Brown. 2007a. Techniques for targeting screening in ulcerative colitis. *Postgraduate Med J* 83 (981): 451–460.

Hurlstone, D. P., R. Kiesslich, M. Thomson, R. Atkinson, and S. S. Cross. 2008b. Confocal chromoscopic endomicroscopy is superior to chromoscopy alone for the detection and characterisation of intraepithelial neoplasia in chronic ulcerative colitis. *Gut* 57 (2): 196–204.

Hurlstone, D. P., and D. S. Sanders. 2006. Recent advances in chromoscopic colonoscopy and endomicroscopy. *Curr Gastroenterol Rep* 8 (5): 409–415.

Hurlstone, D. P., D. S. Sanders, and M. Thomson. 2007b. Detection and treatment of early flat and depressed colorectal cancer using high-magnification chromoscopic colonoscopy: A change in paradigm for Western endoscopists? *Dig Dis Sci* 52 (6): 1387–1393.

Hurlstone, D. P., M. Thomson, S. Brown, N. Tiffin, S. S. Cross, and M. D. Hunter. 2007c. Confocal endomicroscopy in ulcerative colitis: Differentiating dysplasia associated lesional mass and adenoma-like mass. *Clin Gastroenterol Hepatol* 5 (10): 1235–1241.

Hurlstone, D. P., N. Tiffin, S. R. Brown, W. Baraza, M. Thomson, and S. S. Cross. 2008c. *In vivo* confocal laser scanning chromo-endomicroscopy of colorectal neoplasia: Changing the technological paradigm. *Histopathology* 52 (4): 417–426.

Inoue, H., T. Igari, T. Nishikage, K. Ami, T. Yoshida, and T. Iwai. 2000. A novel method of virtual histopathology using laser-scanning confocal microscopy in-vitro with untreated fresh specimens from the gastrointestinal mucosa. *Endoscopy* 32 (6): 439–443.

Just, T., E. Srur, O. Stachs, and H. W. Pau. 2009. Volumetry of human taste buds using laser scanning microscopy. *J Laryngol Otol* 123 (10): 1125–1130.

Khanavkar, B., F. Gnudi, A. Muti et al. 1998. Basic principles of LIFE—autofluorescence bronchoscopy: Results of 194 examinations in comparison with standard procedures for early detection of bronchial carcinoma—overview. *Pneumologie* 52 (2): 71–76.

Kiesslich, R., J. Burg, M. Vieth et al. 2004. Confocal laser endoscopy for diagnosing intraepithelial neoplasias and colorectal cancer *in vivo*. *Gastroenterology* 127 (3): 706–713.

Kiesslich, R., and M. I. Canto. 2009. Confocal laser endomicroscopy. *Gastrointest Endosc Clin N Am* 19 (2): 261–272.

Kiesslich, R., P. R. Galle, and M. F. Neurath. 2007a. Endoscopic surveillance in ulcerative colitis: Smart biopsies do it better. *Gastroenterology* 133 (3): 742–745.

Kiesslich, R., M. Goetz, E. M. Angus et al. 2007b. Identification of epithelial gaps in human small and large intestine by confocal endomicroscopy. *Gastroenterology* 133 (6): 1769–1778.

Kiesslich, R., M. Goetz, J. Burg et al. 2005a. Diagnosing Helicobacter pylori *in vivo* by confocal laser endoscopy. *Gastroenterology* 128 (7): 2119–2123.

Kiesslich, R., M. Goetz, K. Lammersdorf et al. 2007c. Chromoscopy-guided endomicroscopy increases the diagnostic yield of intraepithelial neoplasia in ulcerative colitis. *Gastroenterology* 132 (3): 874–882.

Kiesslich, R., M. Goetz, and M. F. Neurath. 2008a. Confocal laser endomicroscopy for gastrointestinal diseases. *Gastrointest Endosc Clin N Am* 18 (3): 451–466, viii.

Kiesslich, R., M. Goetz, and M. F. Neurath. 2008b. Virtual histology. *Best Pract Res Clin Gastroenterol* 22 (5): 883–897.

Kiesslich, R., M. Goetz, M. Vieth, P. R. Galle, and M. F. Neurath. 2005b. Confocal laser endomicroscopy. *Gastrointest Endosc Clin N Am* 15 (4): 715–731.

Kiesslich, R., M. Goetz, M. Vieth, P. R. Galle, and M. F. Neurath. 2007d. Technology insight: Confocal laser endoscopy for *in vivo* diagnosis of colorectal cancer. *Nat Clin Pract Oncol* 4 (8): 480–490.

Kiesslich, R., L. Gossner, M. Goetz et al. 2006a. *In vivo* histology of Barrett's esophagus and associated neoplasia by confocal laser endomicroscopy. *Clin Gastroenterol Hepatol* 4 (8): 979–987.

Kiesslich, R., M. Gotz, M. F. Neurath, and P. R. Galle. 2006b. Endomicroscopy—technology with future. *Internist (Berl)* 47 (1): 8–17.

Kiesslich, R., M. Gotz, M. F. Neurath, and R. R. Galle. 2006c. The role of new diagnostic procedures in internal medicine. *Internist* 47 (1): 8–17.

Kiesslich, R., A. Hoffman, M. Goetz et al. 2006d. *In vivo* diagnosis of collagenous colitis by confocal endomicroscopy. *Gut* 55 (4): 591–592.

Kiesslich, R., A. Hoffman, and M. F. Neurath. 2006e. Colonoscopy, tumors, and inflammatory bowel disease—new diagnostic methods. *Endoscopy* 38 (1): 5–10.

Kiesslich, R., and M. F. Neurath. 2005c. Endoscopic confocal imaging. *Clin Gastroenterol Hepatol* 3 (7 Suppl. 1): S58–60.

Kiesslich, R., and M. F. Neurath. 2005d. Endoscopic detection of early lower gastrointestinal cancer. *Best Pract Res Clin Gastroenterol* 19 (6): 941–961.

Kiesslich, R., and M. F. Neurath. 2006f. Chromoendoscopy and other novel imaging techniques. *Gastroenterol Clin North America* 35 (3): 605–619.

Kiesslich, R., and M. F. Neurath. 2007e. Endomicroscopy is born—do we still need the pathologist? *Gastrointest Endosc* 66 (1): 150–153.

Kiesslich, R., and M. F. Neurath. 2007f. Magnifying chromoendoscopy: Effective diagnostic tool for screening colonoscopy. *J Gastroenterol Hepatol* 22 (11): 1700–1701.

Kiesslich, R., and M. F. Neurath. 2007g. Screening and early diagnosis of colorectal cancer. *Acta Endoscopica* 37 (2): 207–229.

Knittel, J., L. Schnieder, G. Buess, B. Messerschmidt, and T. Possner. 2001. Endoscope-compatible confocal microscope using a gradient index-lens system. *Opt Commun* 188 (5–6): 267–273.

Kopacova, M., J. Bures, J. Osterreicher et al. 2009. Confocal laser endomicroscopy in experimental pigs: Methods of *ex vivo* imaging. *Cas Lek Cesk* 148 (6): 249–253.

Lane, P. M., A. L. P. Dlugan, R. Richards-Kortum, and C. E. MacAulay. 2000. Fiber-optic confocal microscopy using a spatial light modulator. *Opt Lett* 25 (24): 1780–1782.

Lane, P. M., T. Gilhuly, P. Whitehead et al. 2006. Simple device for the direct visualization of oral-cavity tissue fluorescence. *J Biomed Opt* 11 (2): 024006.

Lane, P. M., S. Lam, A. McWilliams, J. C. Leriche, M. W. Anderson, and C. E. Macaulay. 2009a. Confocal fluorescence microendoscopy of bronchial epithelium. *J Biomed Opt* 14 (2): 024008.

Lane, P. M., R. P. Elliott, and C. E. MacAulay. 2003. Confocal microendoscopy with chromatic sectioning. Paper read at *Proceedings of the SPIE—The International Society for Optical Engineering, Spectral Imaging: Instrumentation, Applications, and Analysis II*, 2003, at San Jose, CA.

Lane, P. M. 2009b. Reflection-contrast limit of fiber-optic image guides. *J Biomed Opt* 14 (6): 064028.

Lane, P. M. 2009c. Terminal reflections in fiber-optic image guides. *Appl Opt* 48 (30): 5802–5810.

Leong, R. W. L., N. Q. Nguyen, C. G. Meredith et al. 2008. *In vivo* confocal endomicroscopy in the diagnosis and evaluation of celiac disease. *Gastroenterology* 135 (6): 1870–1876.

Leung, K. K., D. Maru, S. Abraham, W. L. Hofstetter, R. Mehran, and S. Anandasabapathy. 2009. Optical EMR: Confocal endomicroscopy-targeted EMR of focal high-grade dysplasia in Barrett's esophagus. *Gastrointest Endosc* 69 (1): 170–172.

Liu, H., Y. Q. Li, T. Yu et al. 2008. Gastroenterology—confocal endomicroscopy for *in vivo* detection of microvascular architecture in normal and malignant lesions of upper gastrointestinal tract. *J Gastroenterol Hepatol* 23 (1): 56–61.

Liu, H., Y. Q. Li, T. Yu et al. 2009. Confocal laser endomicroscopy for superficial esophageal squamous cell carcinoma. *Endoscopy* 41 (2): 99–106.

MacAulay, C., P. Lane, and R. Richards-Kortum. 2004. *In vivo* pathology: Microendoscopy as a new endoscopic imaging modality. *Gastrointest Endosc Clin North America* 14 (3): 595–620.

Maitland, K. C., A. M. Gillenwater, M. D. Williams, A. K. El-Naggar, M. R. Descour, and R. R. Richards-Kortum. 2008. *In vivo* imaging of oral neoplasia using a miniaturized fiber optic confocal reflectance microscope. *Oral Oncol* 44 (11): 1059–1066.

Meining, A., M. Bajbouj, and R. M. Schmid. 2007a. Confocal fluorescence microscopy for detection of gastric angiodysplasia. *Endoscopy* 39: E145–E145.

Meining, A., M. Bajbouj, S. von Delius, and C. Prinz. 2007b. Confocal laser scanning microscopy for *in vivo* histopathology of the gastrointestinal tract. *Arab J Gastroenterol* 10 (1): 1–4.

Meining, A., E. Frimberger, V. Becker et al. 2008a. Detection of cholangiocarcinoma *in vivo* using miniprobe-based confocal fluorescence microscopy. *Clin Gastroenterol Hepatol* 6 (9): 1057–1060.

Meining, A., V. Phillip, J. Gaa, C. Prinz, and R. M. Schmid. 2009. Pancreaticoscopy with miniprobe-based confocal laser-scanning microscopy of an intraductal papillary mucinous neoplasm (with video). *Gastrointest Endosc* 69 (6): 1178–1180.

Meining, A., D. Saur, M. Bajbouj et al. 2007c. *In vivo* histopathology for detection of gastro-intestinal neoplasia with a portable, Confocal miniprobe: An examiner blinded analysis. *Clin Gastroenterol Hepatol* 5 (11): 1261–1267.

Meining, A., S. Schwendy, V. Becker, R. M. Schmid, and C. Prinz. 2007d. *In vivo* histopathology of lymphocytic colitis. *Gastrointest Endosc* 66 (2): 398–399, discussion 400.

Meining, A., and M. B. Wallace. 2008b. Endoscopic imaging of angiogenesis *in vivo*. *Gastroenterology* 134 (4): 915–918.

Miehlke, S., A. Morgner, D. Aust, A. Madisch, M. Vieth, and G. Baretton. 2007. Combined use of narrow-band imaging magnification endoscopy and miniprobe confocal laser micros-copy in neoplastic Barrett's esophagus. *Endoscopy* 39 (Suppl. 1): E316.

Monkemuller, K., H. Neumann, and L. C. Fry. 2009. Endoscopic examination of the small bowel: From standard white light to confocal endomicroscopy. *Clin Gastroenterol Hepatol* 7 (2): e11–2.

Morgner, A., M. Stolte, and S. Miehlke. 2007. Visualization of lymphoepithelial lesions in gastric mucosa-associated lymphoid tissue-type lymphoma by miniprobe confocal laser microscopy. *Clin Gastroenterol Hepatol* 5 (9): e37.

Nathanson, M. H. 2004. Confocal colonoscopy: More than skin deep. *Gastroenterology* 127 (3): 987–989.

Nguyen, N. Q., and R. W. L. Leong. 2008. Current application of confocal endomicroscopy in gastrointestinal disorders. *J. Gastroenterol. Hepatol.*

Odagi, I., T. Kato, H. Imazu, M. Kaise, S. Omar, and H. Tajiri. 2007. Examination of normal intestine using confocal endomicroscopy. *J Gastroenterol Hepatol* 22 (5): 658–662.

Palcic, B., S. Lam, J. Hung, and C. Macaulay. 1991. Detection and localization of early lung-cancer by imaging techniques. *Chest* 99 (3): 742–743.

Papworth, G. D., P. M. Delaney, L. J. Bussau, L. T. Vo, and R. G. King. 1998. *in vivo* fibre optic confocal imaging of microvasculature and nerves in the rat vas deferens and colon. *J Anat* 192 (Pt. 4): 489–495.

Paris, S., K. Bitter, H. Renz, W. Hopfenmuller, and H. Meyer-Lueckel. 2009. Validation of two dual fluorescence techniques for confocal microscopic visualization of resin penetration into enamel caries lesions. *Microsc Res Tech* 72 (7): 489–494.

Partridge, M., G. Emilion, S. Pateromichelakis, R. A'Hern, E. Phillips, and J. Langdon. 1998. Allelic imbalance at chromosomal loci implicated in the pathogenesis of oral precan-cer, cumulative loss and its relationship with progression to cancer. *Oral Oncol* 34 (2): 77–83.

Pavlova, I., M. Williams, A. El-Naggar, R. Richards-Kortum, and A. Gillenwater. 2008. Understanding the biological basis of autofluorescence imaging for oral cancer detec-tion: High-resolution fluorescence microscopy in viable tissue. *Clin Cancer Res* 14 (8): 2396–2404.

Pech, O., T. Rabenstein, H. Manner et al. 2008. Confocal laser endomicroscopy for *in vivo* diagnosis of early squamous cell carcinoma in the esophagus. *Clin Gastroenterol Hepatol* 6 (1): 89–94.

Pohl, H., T. Rosch, M. Vieth et al. 2008. Miniprobe confocal laser microscopy for the detection of invisible neoplasia in patients with Barrett's oesophagus. *Gut* 57 (12): 1648–1653.

Polglase, A. L., W. J. McLaren, and P. M. Delaney. 2006. Pentax confocal endomicroscope: A novel imaging device for *in vivo* histology of the upper and lower gastrointestinal tract. *Exp Rev Med Dev* 3 (5): 549–556.

Polglase, A. L., W. J. McLaren, S. A. Skinner, R. Kiesslich, M. F. Neurath, and P. M. Delaney. 2005. A fluorescence confocal endomicroscope for *in vivo* microscopy of the upper- and the lower-GI tract. *Gastrointest Endosc* 62 (5): 686–695.

Reddymasu, S. C., and P. Sharma. 2008. Advances in endoscopic imaging of the esophagus. *Gastroenterol Clin North America* 37 (4): 763–774.

Richards-Kortum, R., and M. Follen. 2006. Presentations to the National Institue of Biomedical Imaging and Bioengineering (NIBIB) by RRK and the the National Cancer Institute (NCI) by MF.

Rosin, M. P., X. Cheng, C. Poh et al. 2000. Use of allelic loss to predict malignant risk for low-grade oral epithelial dysplasia. *Clin Cancer Res* 6 (2): 357–362.

Rosin, M. P., W. L. Lam, C. Poh et al. 2002. 3p14 and 9p21 loss is a simple tool for predicting second oral malignancy at previously treated oral cancer sites. *Cancer Res* 62 (22): 6447–6450.

Sanduleanu, S., A. Driessen, W. Hameeteman, W. van Gemert, A. de Bruine, and A. Masclee. 2009. Inflammatory cloacogenic polyp: Diagnostic features by confocal endomicroscopy. *Gastrointest Endosc* 69 (3): 595–598.

Sidorenko, E. I., and P. Sharma. 2004. High-resolution chromoendoscopy in the esophagus. *Gastrointest Endosc Clin N Am* 14 (3): 437–451, vii.

Sokolov, K., J. Aaron, B. Hsu et al. 2003. Optical systems for *in vivo* molecular imaging of cancer. *Technol Cancer Res Treat* 2 (6): 491–504.

Sonn, G. A., S. N. E. Jones, T. V. Tarin et al. 2009a. Optical biopsy of human bladder neoplasia with *in vivo* confocal laser endomicroscopy. *J Urol* 182 (4): 1299–1305.

Sonn, G. A., K. E. Mach, K. Jensen et al. 2009b. Fibered confocal microscopy of bladder tumors: An *ex vivo* study. *J. Endourol* 23 (2): 197–201.

Sung, K. B., C. Liang, M. Descour et al. 2002. Near real time *in vivo* fibre optic confocal microscopy: Subcellular structure resolved. *J. Microscopy-Oxford* 207: 137–145.

Sung, K. B., R. Richards-Kortum, M. Follen, A. Malpica, C. Liang, and M. Descour. 2003. Fiber optic confocal reflectance microscopy: A new real-time technique to view nuclear morphology in cervical squamous epithelium *in vivo*. *Opt Express* 11 (24): 3171–3181.

Tan, J., P. Delaney, and W. J. McLaren. 2007. Confocal endomicroscopy: A novel imaging technique for *in vivo* histology of cervical intraepithelial neoplasia. *Exp Rev Med Dev* 4 (6): 863–871.

Tan, J., M. A. Quinn, J. M. Pyman, P. M. Delaney, and W. J. McLaren. 2009. Detection of cervical intraepithelial neoplasia *in vivo* using confocal endomicroscopy. *BJOG* 116 (12): 1663–1670.

Tearney, G. J., R. H. Webb, and B. E. Bouma. 1998. Spectrally encoded confocal microscopy. *Opt Lett* 23 (15): 1152–1154.

Thiberville, L., S. Moreno-Swirc, T. Vercauteren, E. Peltier, C. Cave, and G. B. Heckly. 2007. *In vivo* imaging of the bronchial wall microstructure using fibered confocal fluorescence microscopy. *Am J Respiratory Crit Care Med* 175 (1): 22–31.

Thiberville, L., M. Salaun, S. Lachkar et al. 2009a. Confocal fluorescence endomicroscopy of the human airways. *Proc Am Thorac Soc* 6 (5): 444–449.

Thiberville, L., M. Salaun, S. Lachkar et al. 2009b. Human *in vivo* fluorescence microimaging of the alveolar ducts and sacs during bronchoscopy. *Eur Respir J* 33 (5): 974–985.

Thong, P. S. P., M. Olivo, K. W. Kho et al. 2007. Laser confocal endomicroscopy as a novel technique for fluorescence diagnostic imaging of the oral cavity. *J Biomed Opt* 12 (1): 014007.

Trovato, C., A. Sonzogni, D. Ravizza et al. 2007. Celiac disease: *In vivo* diagnosis by confocal endomicroscopy. *Gastrointest Endosc* 65 (7): 1096–1099.

Udovich, J. A., N. D. Kirkpatrick, A. Kano, A. Tanbakuchi, U. Utzinger, and A. F. Gmitro. 2008. Spectral background and transmission characteristics of fiber optic imaging bundles. *Appl Opt* 47 (25): 4560–4568.

van de Ven, A. L., K. Adler-Storthz, and R. Richards-Kortum. 2009. Delivery of optical contrast agents using Triton-X100, part 1: Reversible permeabilization of live cells for intracellular labeling. *J Biomed Opt* 14 (2): 021012.

Van den Broek, F. J. C., P. Fockens, and E. Dekker. 2007. Review article: New developments in colonic imaging. *Aliment Pharmacol Ther* 26: 91–99.

Venkatesh, K., M. Cohen, A. Akobeng et al. 2009a. Diagnosis and management of the first reported case of esophageal, gastric, and small-bowel heterotopia in the colon, using confocal laser endomicroscopy. *Endoscopy* 41(Suppl. 2): E58.

Venkatesh, K., M. Cohen, C. Evans et al. 2009b. Feasibility of confocal endomicroscopy in the diagnosis of pediatric gastrointestinal disorders. *World J Gastroenterol* 15 (18): 2214–2219.

Verveer, P. J., Q. S. Hanley, P. W. Verbeek, L. J. Van Vliet, and T. M. Jovin. 1998. Theory of confocal fluorescence imaging in the programmable array microscope (PAM). *J. Microscopy-Oxford* 189: 192–198.

von Delius, S., H. Feussner, D. Wilhelm et al. 2007. Transgastric *in vivo* histology in the peritoneal cavity using miniprobe-based confocal fluorescence microscopy in an acute porcine model. *Endoscopy* 39 (5): 407–411.

Wallace, M. B., and P. Fockens. 2009. Probe-based confocal laser endomicroscopy. *Gastroenterology* 136 (5): 1509–1513.

Wang, T. D. 2005. Confocal microscopy from the bench to the bedside. *Gastrointestinal Endoscopy* 62 (5): 696–697.

Wang, T. D., C. H. Contag, M. J. Mandella, N. Y. Chan, and G. S. Kino. 2003. Dual-axes confocal microscopy with post-objective scanning and low-coherence heterodyne detection. *Opt Lett* 28 (20): 1915–1917.

Wang, T. D., S. Friedland, P. Sahbaie et al. 2007. Functional imaging of colonic mucosa with a fibered confocal microscope for real-time *in vivo* pathology. *Clin Gastroenterol Hepatol* 5 (11): 1300–1305.

Watanabe, O., T. Ando, O. Maeda et al. 2008. Confocal endomicroscopy in patients with ulcerative colitis. *J Gastroenterol Hepatol* 23: S286–S290.

WHO. 2009a. Cancer, Fact sheet No. 297. World Health Organization,.

WHO. 2009b. The top 10 causes of death, Fact sheet No. 310. World Health Organization.

Yeoh, K. G. 2007. How do we improve outcomes for gastric cancer? *J Gastroenterol Hepatol* 22 (7): 970–972.

Zambelli, A., V. Villanacci, E. Buscarini et al. 2008a. Confocal endomicroscopic aspects in Whipple's disease. *Gastrointest Endosc* 68 (2): 373–374.

Zambelli, A., V. Villanacci, E. Buscarini, G. Bassotti, and L. Albarello. 2008b. Collagenous colitis: A case series with confocal laser microscopy and histology correlation. *Endoscopy* 40 (7): 606–608.

Zhang, J. N., Y. Q. Li, Y. A. Zhao et al. 2008. Classification of gastric pit patterns by confocal endomicroscopy. *Gastrointest Endosc* 67 (6): 843–853.

Zhang, J. N., T. Yu, Y. Q. Li, Y. T. Guo, H. Liu, and J. P. Zhang. 2007. Confocal endomicroscopy for differential diagnosis of indented gastric lesions. *Zhonghua Nei Ke Za Zhi* 46 (10): 835–837.

Zhang, L. W., C. Michelsen, X. Cheng, T. Zeng, R. Priddy, and M. P. Rosin. 1997. Molecular analysis of oral lichen planus—A premalignant lesion? *Am J Pathol* 151 (2): 323–327.

Zheng, W., M. Harris, K. W. Kho et al. 2004. Confocal endomicroscopic imaging of normal and neoplastic human tongue tissue using ALA-induced-PPIX fluorescence: A preliminary study. *Oncol Rep* 12 (2): 397–401.

9 Optical Coherence Tomography

Wenxuan Liang, Jessica Mavadia, and Xingde Li
Department of Biomedical Engineering,
Johns Hopkins University, Baltimore, Maryland

CONTENTS

9.1   INTRODUCTION

Optical coherence tomography (OCT) is a recently developed powerful biomedical imaging technology that is capable of real-time, *in situ* cross-sectional visualization of tissue microstructure. OCT is analogous to ultrasound B-mode imaging in that the backscattered light intensity versus the echo time delay (or depth) is measured to form an axial scan (A-mode). Multiple axial scans acquired by translating the incident beam relative to the sample generate a cross-sectional image (B-mode) and the accrual of cross-sectional images forms a three-dimensional volumetric data set.

The axial resolution of OCT ranges from 1–15 μm. It is determined by the coherence length of the light source, which is inversely proportional to the source spectrum bandwidth. The imaging depth is governed by the optical attenuation of the sample, and often, an imaging penetration depth of 2–3 mm can be achieved in most highly scattering biological tissues when using a near infrared (NIR) light source from 800–1300 nm. Thus, OCT fills a gap between the scales of high-frequency ultrasound and confocal microscopy, providing much finer resolution than the former as well as deeper penetration than the latter (Fujimoto and Drexler 2008).

The high resolution and the inherently noninvasive nature has made OCT very attractive and promising in both biomedical research and clinical applications. Furthermore, as a fiber-optic technology, OCT systems can be compact and portable. Flexible fiber-optic imaging probes allow the integration of OCT with a wide range of medical instruments such as endoscopes, catheters, and laparoscopes etc., enabling imaging of internal organs or lumens with unprecedented resolution. In addition to structural imaging, OCT is also able to extract functional information from deeper layers of the tissue. Several extensions of OCT such as Doppler flow, polarization sensitive, and spectroscopic OCT have also been demonstrated. OCT is becoming a clinically viable imaging modality for early detection of diseases, precision guidance of surgical interventions, and monitoring of treatment effects.

In the following sections, we will summarize the operating principles of OCT, including the low-coherence interferometry theory (for both time domain and Fourier domain OCT), practical considerations of system design (light source, optical design, detectors, probes), and introduce various extensions of OCT (polarization sensitive, spectroscopic, and Doppler OCT). Finally, we will discuss several applications of OCT in biomedicine.

9.2   PRINCIPLES OF OCT AND LOW COHERENCE INTERFEROMETRY

Though analogous to ultrasound, one major difference between OCT and ultrasound is the use of near infrared light instead of sound in OCT. The echo time delay between two points 15 μm apart is as small as 100 fs. Such a short time lies beyond the direct measurement capacity provided by modern electronics, so low coherence interferometry (LCI) is employed. LCI, also known as *white light interferometry,*

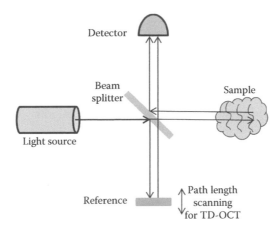

FIGURE 9.1 Schematic of a typical TD-OCT system.

was widely used in photonics to measure optical echoes and backscattering in optical fibers and waveguide devices (Takada et al. 1987; Youngquist, Carr, and Davies 1987; Gilgen et al. 1989) before its first applications in biomedicine (Fercher, Mengedoht, and Werner 1988; Huang et al. 1991). In the following we will elucidate the principles of LCI for both time domain and Fourier domain OCT.

The schematic of a typical OCT system (either time domain or Fourier domain) is based on a fiber-optically implemented Michelson interferometer, as illustrated in Figure 9.1. Low coherence light travels from the source to the beamsplitter, and then is guided to the reference arm and sample arm, respectively. Backreflected waves from the reference arm and the sample arm are combined at the beamsplitter and delivered to the detector, where the intensity of the superposed (interference) fringe signal is measured.

Through our analysis, without loss of generality we would assume the optical path length from beamsplitter to light source is equal to that from beamsplitter to the detector. And for the sake of simplicity, we would further assume an ideal 50/50 beamsplitter of which the splitting ratio is wavelength-independent, and the power loss of light due to transmission through and reflection on the beamsplitter is ignored (with the final results affected only by some constant factor).

9.2.1 TIME DOMAIN OCT

For our theoretical analysis, we could first assume a scalar light wave with angular frequency $\omega = 2\pi f$. Denoting the light speed in vacuum by c, we have the vacuum wavelength $\lambda = c/f$ and the propagation wavenumber $k = 2\pi/\lambda$.

Using the convention that the physical electric (and magnetic) field is the real part of a complex quantity, we write the electric field of the source light as $E_0(k)e^{-j\omega t}$, where $E_0(k)$ is the complex amplitude. Thus at a position where the optical path length measured from the light source is z, the electric field can be written as $E_0(k)e^{-j(kz-\omega t)}$ with the understanding that t is counted from some arbitrarily-selected initial time point for everywhere in space.

The sample under interrogation could be characterized by an electric field reflectivity profile $r_S(z)$ with z being the optical path length measured from light source. Note that we use optical path length to temporarily ignore the possibly varying index of refraction of the sample. Taking into account the round trip traveled by light (we already assumed the optical path length from beamsplitter to light source is the same as to the detector), the electric field of the light leaving the light source with wavenumber k is observed at the detector as the following integral:

$$E_S(k,t) = E_0(k)\int r_S(z)e^{j(2kz-\omega t)}\,dz \tag{9.1}$$

Note that as mentioned before, we already ignored those constant factors resulting from transmission through and reflection on the beamsplitter.

For time domain OCT (TD-OCT), we could imagine the path length scanning in the reference arm as translating a mirror with speed v toward the beamsplitter. Assume that at the initial time $t = 0$, the optical path length measured from the source to the mirror is z_R. With the mirror scanning, the light coming out of the light source with wavenumber k is observed at the mirror with the following electric field:

$$E_{mirror}(k,t) = E_0(k)e^{j(k(z_R-vt)-\omega t)} = E_0(k)e^{j(kz_R-(kv+\omega)t)}$$

Note that angular frequency of the light wave received by the mirror has changed to $kv + \omega$; thus, its wavenumber is no longer k. However, we still use notation k in $E_{mirror}(k, t)$ to indicate the wavenumber of its "original" light emitted by the light source. The advantage of such nomenclature is the ease of tracking its complex field amplitude, that is, $E_0(k)$, which is crucial for analyzing the field of the interference signal later.

So the light returning from the mirror has a different wavenumber $k' = (kv + \omega)/c = k(1 + v/c)$, and the optical path length from the mirror to the detector is $z_R - vt$, thus the light field observed at the detector could be written as

$$E_R(k,t) = r_R E_0(k)e^{j(kz_R-(kv+\omega)t+k'(z_R-vt))} \approx r_R E_0(k)e^{j(2kz_R-(2kv+\omega)t)} \tag{9.2}$$

Here, r_R denotes the electric field reflectivity of a mirror in the reference arm, and the approximation stems from that fact that the scanning speed $v \ll c$. Note that we again use k in $E_R(k, t)$ to indicate its original wavenumber, though the actual angular frequency of the light wave observed at the detector has changed to $2kv + \omega$ due to the Doppler Effect. Also note that throughout the analysis, time t is counted relative to the same initial point for both the sample and reference arm.

The electric field of light observed at the detector is then the sum of the reflected signals from the sample and reference arms. Summing Equation (9.1) and Equation (9.2), we get

$$E_D(k,t) = E_S(k,t) + E_R(k,t) = E_0(k)A(k,t)\cdot e^{-j\omega t} \tag{9.3}$$

where

$$A(k,t) = r_R e^{j2k(z_R - vt)} + \int r_S(z) e^{j2kz} \, dz$$

represents the amplitude modulation due to the interference between the light backreflected from the mirror in the reference arm and the light backscattered from different depths in the sample arm. Additionally, the total electric field could be obtained by integrating Equation (9.3) over the source spectrum, that is, $E_D(t) = \int E_D(k, t)dk$.

However, in practice the quantity that can be directly measured is the intensity (or irradiance) of light, denoted as $I_D(t)$. Physically, intensity is proportional to the electric field squared averaging over the detector time response time T, that is,

$$I_D(t) = \varepsilon_0 c \langle (\operatorname{Re} E_D(t))^2 \rangle_T \tag{9.4}$$

Here, ε_0 is the vacuum permittivity and $\langle \cdot \rangle_T$ represents averaging over T around time point t with $T \gg 2\pi/\omega$. In the following we will just write $I_D(t) = \langle (\operatorname{Re} E_D(t))^2 \rangle_T$ (i.e., ignoring the constant factor $\varepsilon_0 c$) without causing confusion.

When substituting $E_D(t)$ into Equation (9.4), cross-product terms between $E_D(k, t)$'s featuring different k's can be ignored thanks to the low-coherence light source adopted in OCT imaging. So the total intensity reduced to the integral of $\langle (\operatorname{Re} E_D(k,t))^2 \rangle_T$ over the range of wavenumber k.

The fact that $kv \ll kc = \omega$ implies that $A(k, t)$ stays almost unchanged during the time averaging interval T, thus the product $E_0(k)A(k, t)$ could be regarded as the complex amplitude of the electric field in Equation (9.3). After some algebra we find:

$$\langle (\operatorname{Re} E_D(k,t))^2 \rangle_T = \frac{1}{2} |E_0(k)A(k,t)|^2$$

$$= \frac{1}{2} |E_0(k)|^2 \left[r_R^2 + \iint r_S(a)r_S(b) e^{j2k(a-b)} \, da \, db \right.$$

$$\left. + 2 \int r_R r_S(z) \cos(2k(z_R - vt - z)) dz \right]$$

Here, r_R^2 represents a constant bias component resulting from the reference arm. The double integral comes from autocorrelation of the backscattered signal from the sample arm. Both terms remain unchanged with time, serving as the DC component in the measured fringe signal intensity. The last integral represents the interference between lights backscattered from the reference and the sample arm, which varies with time as an "AC" component of the total interference signal, and is of interest for TD-OCT imaging.

Denoting the power density spectrum of the light source by $S_0(k)$ and understanding that $S_0(k) \propto |E_0(k)|^2$, the AC component of the OCT intensity signal is

$$I_D^{AC}(t) \propto \int S_0(k) \left(\int r_R r_S(z) \cos(2k(z_R - vt - z)) \, dz \right) dk$$

$$\propto \int r_R r_S(z) \left(\int S_0(k) \cos(2k(z_R - vt - z)) \, dk \right) dz \qquad (9.5)$$

$$\propto \int r_R r_S(z) \Psi(2(z_R - vt - z)) \, dz$$

Here, function $\Psi(z)$ is the real part of the Fourier transform of power density spectrum $S_0(k)$, that is,

$$\Psi(z) = \int S_0(k) \cos(kz) \, dk = \text{Re} \left\{ \int S_0(k) e^{-jkz} \, dk \right\} \qquad (9.6)$$

In Equation (9.5) it is manifest that $\Psi(2z)$ acts as the point spread function (PSF) for the TD-OCT system. Thus, it is the spread of $\Psi(2z)$ that determines the axial resolution of TD-OCT.

For simplicity, we consider a low-coherence light source with a Gaussian-shape power density spectrum around a center wavenumber k_0, that is,

$$S_0(k) \propto e^{-4 \ln 2 \cdot (k - k_0)^2 / \Delta_k^2} \qquad (9.7)$$

Here, Δ_k is the full-width-at-half-maximum (FWHM) of the power density spectrum in k-domain. After some algebra one could see that the Fourier transform of $S_0(k)$ features the following form:

$$\hat{S}_0(z) \doteq \int S_0(k) e^{-jkz} \, dk \propto e^{-jk_0 z} \cdot e^{-4 \ln 2 \cdot z^2 / \Delta_z^2} \qquad (9.8)$$

where Δ_z is the FWHM of $\hat{S}_0(z)$ in z-domain and satisfies $\Delta_z \cdot \Delta_k = 8 \ln 2$.

Substituting $\Psi(z) = \text{Re}(\hat{S}_0(z))$ into Equation (9.5), we obtain the final expression for the AC component of the fringe signal:

$$I_D^{AC}(t) \propto \int r_R r_S(z) \cdot e^{-16 \ln 2 \cdot (z_R - vt - z)^2 / \Delta_z^2} \cos(2k_0(z_R - vt - z)) \, dz \qquad (9.9)$$

Equation (9.5) and Equation (9.9) jointly serve as the starting point for OCT signal processing. They reveal that the AC component measured at the detector is a modulated sinusoidal signal with a circular frequency of $2k_0 v = 2\omega_0 v/c$, where $\omega_0 = ck_0$ is the central angular frequency of the light source. The rapidly-decaying exponential term in Equation (9.9) tells that for any given time t, the integral is governed by $r_S(z)$ with $z \approx z_R - vt$, that is, the modulation amplitude $I_D^{AC}(t)$ reflects the profile

$r_S(z_R - vt)$. Such intensity versus depth information composes an axial scan which is analogous to B-mode ultrasonic imaging but provides much higher resolution.

Quantitatively the axial resolution is the FWHM of the point spread function $\Psi(2z)$, that is, one-half of Δ_z as in Equation (9.8). Thus the axial resolution expressed in optical path length, denoted by δ_z, is

$$\delta_Z = \frac{\Delta_z}{2} = \frac{4\ln 2}{\Delta_k} = \frac{2\ln 2}{\pi} \cdot \frac{\lambda_0^2}{\Delta_\lambda} \qquad (9.10)$$

where λ_0 is the central wavelength of the light source. In tissue the spatial resolution is given by δ_z/n where n is the index of refraction. Since the OCT axial resolution is inversely proportional to the spectrum bandwidth of the light source, that is, Δ_k or equivalently $\Delta\lambda$, an ideal light source for OCT should have a broad bandwidth (thus a short temporal coherence length).

The transverse (or lateral) resolution of OCT is the same as in optical microscopy and is determined by the diffraction-limited focused spot size of the imaging beam, which is inversely proportional to the numerical aperture as

$$\delta_x = \frac{4\lambda_0}{\pi} \cdot \frac{f}{d}$$

with d the beam spot size on the objective lens and f the focal length. Again, similar to optical microscopy, the depth of field or the confocal parameter b is given by

$$b = \frac{\pi}{\lambda} \delta_x^2$$

It is noted that there exists a tradeoff between the transverse resolution and the depth of field. Typically, low NA focusing optics is employed in OCT imaging in order to have a larger depth of field and thus maintain a good transverse resolution over a reasonable depth range. If a high transverse resolution is prioritized, high NA focusing optics could be adopted, and the depth of field would decrease quadratically with respect to the transverse resolution and could be made comparable or even smaller than the coherence length. In this case confocal gating is integrated with coherent gating to distinguish signals from a given depth. This *en face* imaging mode is known as optical coherence microscopy (OCM) (Izatt et al. 1996; Aguirre et al. 2003).

Besides what is mentioned above, other requirements for an OCT light source include emission in the near-infrared range (e.g., 800–1500 nm) for deeper tissue penetration, high irradiance power for better SNR and a near-Gaussian spectrum shape to minimize the side-lobe on the interference signal. When used in conjunction with fiber-optic probes, single spatial mode is also required. More details about different kinds of light sources will be expatiated in Section 9.3.1.

9.2.2 FOURIER DOMAIN OCT

The basic schematic of a Fourier domain OCT (FD-OCT) system is the same as TD-OCT except that the reference arm scanning is eliminated and the spectrum of the interference signal is measured directly for axial (or depth) information retrieval via Fourier transform. There are two types of FD-OCT distinguished by their methods of spectrum detection. One type called spectral domain OCT (SD-OCT) employs a detector array (e.g., a high-speed line-scan CCD or CMOS camera) to record the spectral interference signal at all wavelengths simultaneously. The other type known as swept source OCT (SS-OCT) or optical frequency domain imaging (OFDI), utilizes a wavelength-tunable light source and a single detector that captures the interference fringe signal one wavelength at a time. Swept sources have been implemented in several ways such as by mechanical scanning (Yun et al. 2003), via ring lasers (Choma, Hsu, and Izatt 2005), and more recently by Fourier domain mode locking (FDML) (Huber, Wojtkowski, and Fujimoto 2006), which will be further explained in the section of light sources. Both SD- and SS-OCT have been shown to significantly improve the signal-to-noise ratio (SNR) and imaging speed as compared to TD-OCT (Choma et al. 2003; de Boer et al. 2003; Leitgeb, Hitzenberger, and Fercher 2003).

The theoretical analysis for FD-OCT turns out to be almost the same as for TD-OCT. First the absence of path length scanning in the reference arm eliminates the $2kv$ term in Equation (9.2). Thus, the electric field of the backscattered light from reference arm with a wavenumber k is simplified to

$$E_R(k,t) = r_R E_0(k) e^{j(2kz_R - \omega t)} \tag{9.11}$$

With $E_S(k, t)$ unchanged as in Equation (9.1) and $E_D(k, t)$ as in Equation (9.3), the intensity of interference signal directly measured at wavenumber k, that is, $I_D(k) = 1/2 |E_D(k, t)|^2$ now becomes

$$I_D(k) \propto S_0(k) \left\{ r_R^2 + \iint r_S(a) r_S(b) e^{j2k(a-b)} \, da \, db + \int r_R r_S(z) e^{j2k(z-z_R)} \, dz \right.$$
$$\left. + \int r_R r_S(z) e^{-j2k(z-z_R)} \, dz \right\} \tag{9.12}$$

To retrieve depth information, the Fourier transform is applied to the measured $I_D(k)$. With $\hat{S}_0(z)$ still denoting the Fourier transform of $S_0(k)$, the overall Fourier transform of $I_D(k)$, denoted as $\hat{I}_D(z)$, is the sum of transforms of the three terms in Equation (9.12). After some algebra we get

$$\hat{I}_D(2z) \propto r_R^2 \hat{S}_0(2z) + \iint r_S(a) r_S(b) \hat{S}_0(2(z+b-a)) \, da \, db$$
$$+ r_R \int r_S(a) \hat{S}_0(2(z+z_R-a)) \, da + r_R \int r_S(a) \hat{S}_0(2(z-z_R+a)) \, da \tag{9.13}$$

Note that the factor of 2 is introduced deliberately into the equation to facilitate the following analysis. Note that the light source used for the FD-OCT imaging can be modeled in the same way as in Equation (9.7) and Equation (9.8), with a broad bandwidth and a short coherence length Δ_z, which decides the rapid decay of $\hat{S}_0(z)$ off $z = 0$.

First, to visualize how $r_S(z)$ is encoded in the transform result $\hat{I}_D(z)$, we could idealize $\hat{S}_0(z)$ to be Dirac delta function $\delta(z)$. Utilizing the identity $\delta(2z) = 1/2\delta(z)$, from Equation (9.13) we obtain

$$\hat{I}_D(2z) \propto r_R^2\delta(z) + \int r_S(a)r_S(a-z)\,da + r_Rr_S(z_R+z) + r_Rr_S(z_R-z) \qquad (9.14)$$

Physically speaking, the factor of 2 on the left-hand side (LHS) actually stems from the round trip traveled by the light. The right-hand side (RHS) composes of four components. The first term $r_R^2\delta(z)$ represents the reflection from the reference mirror which is often large and becomes an artifact located at the origin. The second term is autocorrelation of the tissue reflectivity profile, which is relatively small due to the low tissue reflectivity compared to r_R. This autocorrelation term also acts as a small artifact around the origin.

The last two terms in Equation (9.14) are mirror images (symmetric with respect to $z = 0$) of each other, and reflectivity profile $r_S(z)$ could be recovered from either one. Note that $r_S(z + z_R)$ is $r_S(z)$ left-translated by r_R, and $r_S(z_R - z)$ is the mirror image of $r_S(z)$, that is, $r_S(-z)$, right-translated by z_R. To prevent overlapping of these two, one could simply make sure that $r_S(z) = 0$ for $z < z_R$. Practically this could be realized by placing the mirror in reference arm closer to the beamsplitter than the first surface of the sample imaged (Hausler and Lindner 1998).

Secondly, to clarify the PSF and thus the axial resolution of FD-OCT imaging, we can place a mirror in the sample arm, and thus the reflectivity profile $r_S(a)$ in Equation (9.13) can be set to a Dirac delta function with an optical path length $z_S > z_R$, that is, $r_S(z) = r_S \cdot \delta(z - z_S)$, and then Equation (9.13) is reduced to

$$\hat{I}_D(2z) \propto (r_R^2 + r_S^2)\hat{S}_0(2z) + r_Rr_S\hat{S}_0(2(z + z_R - z_S)) + r_Rr_S\hat{S}_0(2(z - z_R + z_S)) \qquad (9.15)$$

From Equation (9.8) and considering the source power density spectrum $S_0(k)$ is a real function of wavenumber k, we have $\hat{S}_0^*(2(-z + z_R - z_S)) = \hat{S}_0(2(z - z_R + z_S))$ where $\hat{S}_0^*(z)$ denotes the complex conjugate of $\hat{S}_0(z)$. Equation (9.15) can then be rewritten as

$$\hat{I}_D(2z) \propto (r_R^2 + r_S^2)\hat{S}_0(2z) + r_Rr_S\hat{S}_0(2(z + z_R - z_S)) + r_Rr_S\hat{S}_0^*(2(-z + z_R - z_S)) \qquad (9.16)$$

We clearly see that the last two terms in Equation (9.16) are conjugate symmetric to each other, and thus their modulus profiles are symmetric to each other with respect to $z = 0$. Moreover, the rapid decay of $\hat{S}_0(z)$ off $z = 0$ for a broadband light source governs that these two terms concentrates around $\pm(z_S - z_R)$, respectively. From the above analysis, we find that only one of the last two terms in Equation (9.16) is needed to estimate the reflectivity profile of the sample arm r_s. As shown explicitly in Equation (9.16), $\hat{S}_0(2z)$ serves as the PSF of the FD-OCT system. Since the FWHM of $\hat{S}_0(2z)$ is determined by the same envelope the same as its real part $\Psi(2z)$, the axial resolution of FD-OCT system is given by the same identity as in Equation (9.10).

One important practical concern stems from the fact that discrete Fourier transform (DFT) has to be employed since $I_D(k)$ can be sampled only at a discrete set of k values, and so the penetration depth is determined by the sampling density in k-space. According to the Nyquist sampling theory, the sampling interval in k-space, denoted by τ_k, decides that the maximal discernible range in z-domain to be $|z| < \pi/\tau_k$. The factor of 2 on the LHS of Equation (9.14) determines that, a physical penetration depth z relative to the mirror in the reference arm corresponds to $2z$ in the transformed z-domain. Since only one side of $\hat{I}_D(z)$ is useful, the maximal penetration depth for ordinary FD-OCT imaging is $\pi/(2\tau_k)$ in free space, and $\pi/(2n\tau_k)$ in tissue with a group index of refraction n. In SS-OCT system τ_k is decided by the instantaneous line-shape of the swept laser source, while in SD-OCT system it is determined by the spectral resolution of the spectrometer employed.

Our final remark deals with the mirror image which occupies half of the z-domain and provides nothing but redundant information. To fully utilize the z-domain, several phase-shifted copies of the interference signal can be taken and combined linearly via appropriate complex coefficients. Then Fourier transforming the combinatorial signal eliminates the duplicated mirror image and doubles the useful z-domain and thus penetration depth. Such techniques have been coined as complex full-range Fourier domain OCT and have been applied to both SD- and SS-OCT systems via various mechanisms (Wojtkowski et al. 2002; Leitgeb, Hitzenberger et al. 2003; Yun et al. 2004; Targowski et al. 2005; Davis, Choma, and Izatt 2005; Sarunic et al. 2005; Bachmann, Leitgeb, and Lasser 2006; Yasuno et al. 2006; Wang 2007; Baumann et al. 2007).

9.3 OCT SYSTEM COMPONENTS

9.3.1 Light Sources

Several properties govern light sources appropriate for OCT. These properties include wavelength, bandwidth, power, stability, and the shape of the spectrum (Bouma and Tearney 2002; Brezinski 2006). Several sources commonly used for OCT are semiconductor continuous wave (CW) sources such as superluminescent diodes or doped fiber sources, supercontinuum sources, wavelength scanning or swept sources such as Fourier domain mode locking fiber lasers, and solid-state femtosecond lasers. In the following sections, we attempt to briefly discuss these properties and will describe a variety of sources that have been used for OCT to date.

9.3.1.1 Wavelength

As seen in Equation (9.10), the axial resolution of OCT is directly related to the source spectral bandwidth $\Delta\lambda$ (FWHM) and the central λ_0, thus sources with longer central wavelengths require a larger bandwidth for a given resolution. It is also noted that the signal-to-noise ratio in an OCT system is related to the available (or permissible) source power, that is,

$$\frac{SNR v_S}{P_S L_C} = const. \tag{9.17}$$

Here SNR represents the signal-to-noise ratio or the sensitivity, P_S is the maximal power, v_S is a parameter proportional to the scanning rate or imaging speed, and L_C represents the source coherence length. The importance of this equation lies in the fact that increasing the optical power (within the bounds of the maximum permissible exposure limit) can help increase the image acquisition rate or resolution while the sensitivity remains unchanged (Bouma and Tearney 2002).

In addition, the optical properties of tissue (i.e., scattering and absorption) are wavelength dependent, which in turn govern the imaging penetration depth. In the NIR range (800–1300 nm), tissue attenuation is dominated by scattering which decreases versus wavelength, and thus the imaging depth generally increases with wavelength. Furthermore, OCT imaging contrast is also highly correlated with imaging wavelength. Generally speaking, scattering contrast is more pronounced and monotonically decreases with wavelength unless strong absorbers are present within the source spectral bandwidth.

Currently, the most common wavelengths for OCT imaging are centered around 800 nm, 1060 nm, and 1300 nm. Good tissue contrast is provided by 800 nm but with limited penetration in highly scattering tissue (i.e., about 0.5–0.8 mm). One of the advantages is that water absorption is extremely low at this wavelength which has made it the most popular wavelength for ophthalmic applications, particularly for posterior eye imaging (such as the retina). In comparison, tissue scattering is lower at 1300 nm, while absorption increases; the total optical attenuation (i.e., the sum of tissue scattering and absorption) remains low and thus results in a better imaging penetration depth in scattering-dominant tissues. The imaging contrast at 1300 nm, however, is generally reduced compared to 800 nm. Note that 1300 nm would not be appropriate for posterior eye imaging because of the high water absorption and the long path-length of the eyeball. Between 800 nm and 1300 nm, 1060 nm is a good compromise with a good tradeoff between imaging depth and imaging contrast. In addition, tissue dispersion is low at 1060 nm which is another advantage of using this wavelength.

9.3.1.2 Spectral Shape

The shape of the source is important for preserving some important characteristics for OCT such as the axial PSF and the shape of the PSF (e.g., the side lobes). As discussed previously, a broad spectrum is required to produce a superb OCT axial resolution. Additionally, low source intensity fluctuation is also critical for minimizing the noise in the OCT signal. Intensity fluctuation generally results from variations in the source current and mechanical instability of the OCT interferometer. Fortunately, the excessive noise associated with the intensity fluctuation, as commonly observed in a supercontinuum light source, can be suppressed effectively using a dual-balanced detection scheme (Hartl et al. 2001).

9.3.1.3 Semiconductor Sources

Semiconductor light sources can be compact, reliable, quiet and broadband, and are often high-quality sources for OCT imaging (Chan et al. 2006). Among semiconductor sources, superluminescence diodes (SLD) are most commonly used for their smooth, broad and stable spectra. SLDs are available with a range of central wavelengths from 600–1600 nm. With technological advances SLDs now offer much broader spectral bandwidths as well as higher optical output powers.

Semiconductor optical amplifiers (SOA) are also commonly used as light sources for OCT. SOAs produce spontaneous and stimulated emission by using a current to inject electrons and causing electrons to combine with holes (Connelly 2002). SOAs are used in several configurations such as Fourier domain mode-locking and ring lasers. SOAs have many advantages in terms of compactness, high power and broad bandwidth (50–100 nm), achieving excellent OCT imaging performance.

9.3.1.4 Supercontinuum Sources

Continuum generation generally refers to processes that generate a broad spectrum. For OCT imaging, a broad spectrum results in a better axial resolution. It has been shown that a supercontinuum can be generated by pumping a microstructured fiber with a Kerr-lens mode locked femtosecond Ti:Sapphire laser, resulting in a 2.5 µm axial resolution (Hartl et al. 2001). The advantages of a supercontinuum source are its broad spectrum and high power; while the challenges are often the nonideal spectrum shape and the relatively high noise associated with the nonlinear continuum generation process, they can be suppressed to a certain degree with dual-balanced detection.

9.3.1.5 Femtosecond Sources

Femtosecond light sources can provide a large bandwidth and thus superb resolution. The axial resolution of OCT is equal to the transform-limited temporal pulse width τ times the speed of light in the medium v, that is, $\tau \cdot v$ or $\tau \cdot c/n$ where n is the refractive index of the medium. Using a Kerr-lens mode locked Ti:Sapphire (Ti:Al$_2$O$_3$) laser, a bandwidth of 350 nm was obtained around 800 nm, resulting in an ultrahigh OCT axial resolution and a large spectral range for spectroscopic OCT in a region sensitive to oxy- and deoxyhemoglobin, making it feasible to perform functional imaging in addition to conventional structural imaging (Morgner et al. 2000). In addition to femtosecond Ti:Sapphire lasers, it has been shown that ultrahigh resolution OCT can be performed using a femtosecond Cr^{4+}:Forsterite (Cr^{4+}Mg$_2$SiO$_4$) laser with a central wavelength around 1.28 µm and a spectral bandwidth about 250 nm (Chudoba et al. 2001). Compared to supercontinuum light sources, solid-state mode-locked femtosecond lasers offer not only a broad spectrum but also superb stability for achieving an excellent OCT signal-to-noise ratio.

9.3.1.6 Swept Sources

Swept sources present a novel way to perform OCT imaging with high speed and an excellent signal-to-noise ratio. In essence, the idea is to filter and scan a broadband source or tune the length of the laser cavity to produce a spectral range which will allow imaging. Swept sources have been implemented in several ways such as through ring lasers (Choma, Hsu, and Izatt 2005), Fourier domain mode locking (FDML) fiber lasers (Huber, Wojtkowski, and Fujimoto 2006), and mechanical scanning (Golubovic et al. 1997; Yun et al. 2003).

Among swept laser sources, the FDML fiber laser is of particular interest due to its excellent performance in terms of sweeping speed, output power, and stability. Currently, FDML enables imaging at a speeds higher than 1G axial scans per second (Klein et al. 2011). An FDML fiber laser is essentially a ring laser using an SOA as the gain medium and a tunable filter for wavelength selection (Huber et al. 2005). Generally speaking, each time the cavity length is altered a new wavelength begins to be built up to a sufficient power. Stable operation can be reached when the looping frequency at any given wavelength within the ring cavity is synchronized with the tuning frequency of the Fabry-Perot filter.

Recently a self-starting, self-regulating FDML fiber laser with an active feedback mechanism has been developed to provide an automatic startup without a priori knowledge (Murari et al. 2011). This robust configuration provides a wide FWHM bandwidth of 140 nm centered around 1305 nm with a stable spectrum. Additionally, the ability to automatically readjust the filter tuning frequency to match any potential variation in the fiber-optic cavity length (e.g., owing to temperature fluctuation), makes it an ideal choice for commercial applications where manual intervention would be inconvenient.

9.3.2 SIGNAL DETECTION

SD-OCT systems employ a spectrometer to separate the spectral interference fringe signal according to the wavelength, and typically data collected from an ordinary spectrometer is nonuniform in the spatial frequency domain (or the wavenumber k-space). In practice, fast Fourier Transform (FFT) is performed to extract the depth-dependent reflectivity in the sample from the measured spectral fringe data, which requires the data to be uniformly distributed in k-space. One straightforward solution is to digitally interpolate the measured data and then resample it uniformly in k-domain, where usually a higher-order polynomial (e.g., cubic) is employed. Another approach is to employ nonuniform fast Fourier transform algorithms (Dutt and Rokhlin 1993; G 1995; Beylkin 1995; Fessler and Sutton 2003) rather than standard FFT, thus eliminating the need for any numerical interpolation. Apart from these computation-intensive software approaches, various hardware-based solutions have been proposed to achieve uniform sampling in k-domain, including optical frequency comb (Bajraszewski et al. 2008), translation-slit-based wavelength filter (Jeon et al. 2011), and linear-wavenumber spectrometer which features a specifically designed compensation prism (Hu and Rollins 2007; Gelikonov, Gelikonov, and Shilyagin 2009).

For SS-OCT, signals can be detected with PIN photodiodes. PIN photodiodes have a faster response time and thus are ideal for use in high frequency applications. Currently, however, balanced detectors are more popular. Balanced detectors have two photodiodes to cancel out any common noise or DC components. Similar issues arise about how to convert the sampled data to uniformly distributed data in k-space for efficient signal processing. Besides classical methods via numerical interpolation, there are also hardware solutions to achieve uniform k-space sampling (Eigenwillig et al. 2008; Xi et al. 2010).

9.3.3 ENDOSCOPE

Translating OCT technology into the clinic for imaging of internal organs requires miniature endoscopes or catheters. In general endoscopic imaging probes are either side-viewing or forward-viewing, and each has its respective advantages. Forward-facing endoscopes image tissue in front of the probes, suitable for image-guided surgeries or sample targeting (Liu et al. 2004). Side-viewing probes can be used to image the entire circumference of a lumen providing a cross-sectional image of the lumen (Tearney et al. 1996). Designing forward-facing probes requires consideration of beam scanning mechanisms to image the entire field of view, whereas often, side-viewing probes require mechanical considerations to provide rotational movement. Scanning within the endoscope can be performed via mechanical components at the proximal or distal end of the probe. Among many scanning mechanisms, the ones that are commonly used in a scanning OCT endoscope include a mechanical rotor at the proximal end of the probe (Tearney et al. 1996), and piezoelectric transducers (Liu et al. 2004; Huo et al. 2010) and micro-electro-mechanical systems (MEMS) (Pan, Xie, and Fedder 2001; Zara et al. 2003; Herz et al. 2004; Tran et al. 2004; Wang et al. 2005) at the distal end of the probe. In the following, we will discuss design considerations in forward-viewing, side-viewing, and needle imaging probes.

9.3.3.1 Basic Design Considerations

While there are multitudes of distal end optic designs for endoscopic OCT probes, here we will consider the most basic design, which consists of three main components, a single-mode fiber, a spacer rod, and a GRIN lens as shown in Figure 9.2, which all together control the key imaging beam parameters including the working distance and focused spot size. The fiber guides the light to a spacer rod which will expand the beam before it is focused by a GRIN lens. By adjusting the spacer rod and GRIN lens a focused spot size and desired working distance can be achieved. In the following sections several designs will be discussed in more detail.

9.3.3.2 Side-Viewing Endoscope/Balloon Catheter-Endoscope

Side-viewing probes such as balloon catheters perform circumferential imaging to assess luminal organs. Circumferential scanning can be realized in a variety of ways as mentioned, we will first consider proximal end rotation through a commercially available fiber-optic rotary joint. Rotary joints couple a stationary source fiber with the rotating endoscope. Having chosen the mechanism by which scanning will be

FIGURE 9.2 Basic endoscope design schematic consisting of a fiber, a spacer rod, and GRIN lens.

performed, the next consideration is the design of distal end optics. A major challenge in the distal end optics design lies in obtaining a high resolution at a desired working distance while maintaining a small overall probe diameter to ease the delivery of the catheter to the lumen directly or through a working channel of a standard endoscope.

One such design involves a compound micro lens structure consisting of two GRIN lenses and a spacer rod. The principle behind this design is to first tightly focus the beam emerging from the single-mode fiber to smaller than the mode field diameter of a single-mode fiber through the combination of the glass rod and first GRIN lens, which together form a spot size reducer. Once the spot size has been reduced, the beam passes through the second GRIN lens where it will expand much faster due to the previous reduction in spot size. This design can achieve a long working distance of 9.6 mm with a spot size of ~39 μm suitable for imaging a large lumen such as human esophagus (Fu et al. 2008).

Furthermore, an improvement upon this initial design came from the realization that the focused spot size and working distance can be tuned by the distance from the fiber tip to the GRIN lens and the focusing power (or pitch number) of the GRIN lens. This led to a single GRIN lens based design as shown in Figure 9.3A, which was able to achieve a working distance of ~11.2 mm with an even smaller focused spot size ~21.2 μm (Xi et al. 2009). In general the micro compound lens enables maximum usage of the available numerical aperture (NA) provided by the miniature optics and thus enables a small focused spot size.

In both of these designs the key parameters are the working distance and corresponding resolution and ultimately these are controlled by the object distance and pitch number of the GRIN lens. The object distance is the distance from the fiber tip to the GRIN lens, while the pitch number controls the focusing power of the GRIN lens. The first parameter that should be chosen is the working distance, and this depends upon the application for which the catheter will be used. Having chosen a working distance, the spacer and GRIN lens lengths are tuned to provide an optimal focused spot size. A practical consideration in working with GRIN lenses is to ensure that the beam diameter is no greater than a maximum of ~80% of the GRIN

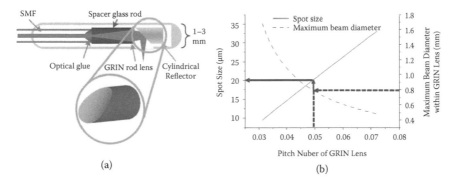

FIGURE 9.3 (A) Distal-end optics for probe design. (B) Graph depicting the relationship between the beam spot size, pitch number of the GRIN lens and the maximum beam diamter within the GRIN lens. (Figure reprinted with permission from Xi, J. et al. 2009. *Opt. Lett.* 34 (13): 1943–1945.)

lens diameter to maximize efficiency, and minimize degradation of the beam profile as a consequence of beam vignetting. Thus, for a given 1-mm-diameter GRIN lens it is necessary to ensure that the maximum beam diameter within the GRIN lens is no more than 0.8 mm. Since the maximum beam diameter within the GRIN lens is related to the pitch number, it is important to calculate the pitch number to obtain the resulting focused spot size. In our case with 0.8 mm as the beam diameter, the pitch number is ~0.05 from the pitch-beam diameter curve, and following the pitch number vertically to the pitch-spot size curve we obtain a focused spot size of ~21 μm as shown in Figure 9.3B (Xi et al. 2009).

Another significant challenge lies in correcting aberrations. In a side-viewing catheter, the imaging beam in many applications needs to pass through a transparent protective plastic tube, which induces a cylindrical lens effect, and consequently causes severe aberrations along the circumferential direction. Recently it has been shown that this generic challenge can be effectively overcome by using an elegant yet simple design as shown in Figure 9.3A. In essence, a cylindrical reflector was used to focus the beam along the circumferential direction, in addition to deflecting the beam by 90°. Such a curved reflector of an appropriate radius of curvature significantly reduced the astigmatism and obtain a near diffraction-limited small focused spot size of ~20 μm for a small catheter with a long working distance ~12 mm (Xi et al. 2009).

Long working distances enable balloon imaging of large luminal organs such as esophagus and one advantage of circumferential imaging is the inherent ability to obtain 3D volumetric images. 3D volumetric images are easily obtained by using spiral scanning, that is, pulling the catheter along its longitudinal direction while the catheter is rotating circumferentially. Figure 9.4 shows a representative 2D snapshot of a circumferential 2D image and a 3D image acquired with a balloon OCT imaging catheter endoscope from a pig esophagus.

Another popular approach to side-viewing catheters is through distal end rotation as opposed to rotating the catheter proximally, and this is done through the use of a

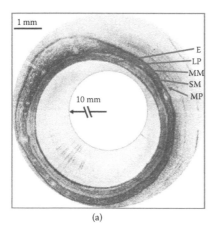
(a)

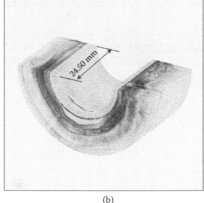
(b)

FIGURE 9.4   (A) 2D circumferential image and (B) 3D volumetric cutaway view of OCT image of a pig esophagus *in vivo*. (Figure reprinted with permission from Xi, J. et al. 2009. *Opt. Lett.* 34 (13): 1943–1945.)

miniature motor. This approach has advantages in that it offers rotational scanning beam uniformity without exerting extra torque on the catheter fiber and thus eliminates the associated birefringence artifacts. There have been several approaches ranging from the use of a DC motor (Herz et al. 2004) to MEMS (Tran et al. 2004) and a squiggle motor (Chang et al. 2011). One of the first designs based on a micromotor had an outer diameter of 4.8 mm and was enclosed in a 5 mm diameter plastic sheath. The micromotor at the distal end was used to rotate a 45° rod mirror to scan the imaging beam circumferentially. Since this device was fabricated, many modifications have led to smaller versions. Through the use of a 1.9 mm diameter MEMS motor, which allowed for an overall rigid diameter of 2.4 mm, a smaller distal scanning catheter endoscope probe was fabricated; however, this probe did not allow for adjustment of the focus (Tran et al. 2004). Recently, another endoscope probe has been designed based upon a squiggle motor (New Scale Technologies), which allows for the fiber to remain stationary while a mirror deflecting the beam to the sample is rotated by the motor (Chang et al. 2011). This design has several advantages since it is able to provide 360° unobstructed view, unlike distally placed motors whose electrical wires will partially block the imaging view. Additionally, the need for a fiber optic rotary joint is obviated, thus minimizing the loss of imaging signal.

9.3.3.3 Forward-Viewing Probe

Forward-viewing probes utilize a different design concept than discussed above where beam scanning is typically not performed through a fiber-optic rotary joint or motor. One approach to perform beam scanning in a forward-viewing endoscope, is through a tubular ceramic piezoelectric (lead zirconate titanate—PZT) actuator. The outer surface of the PZT actuator is separated into four quadrants forming two orthogonal pairs of electrodes (X and Y, respectively). Maximum lateral scanning by the fiber tip is achieved when the PZT drive frequency is equal to the mechanical resonant frequency of the fiber-optic cantilever, which is given by:

$$f = \left(\frac{\beta}{4\pi} \right) \left(\frac{E}{\rho} \right)^{1/2} \left(\frac{R}{L^2} \right)$$ (9.18)

where L is the length of the cantilever, R the radius of the cantilever, E Young's modulus, ρ the mass density, and β a constant related to the boundary conditions and vibration mode number (Liu et al. 2004). The sweeping fiber tip will be imaged to the sample through a miniature lens (such as a GRIN lens), and the distance from the fiber tip to the lens can be used to adjust the focused spot size as well as the working distance of the probe. The schematic of a forward-viewing endoscope probe is shown in Figure 9.5.

In this design beam scanning is performed by applying various waveforms to the electrode pairs to displace the PZT tube and fiber tip. For example, a line scan can be performed by applying sinusoidal waves of opposite polarities to one pair of electrodes such that they will move transversely with respect to the corresponding quadrants (Liu et al. 2004). By controlling each set of electrodes it is possible to displace the fiber cantilever and form spiral or Lissajous scanning patterns.

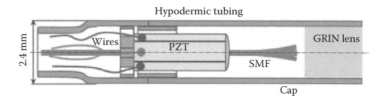

FIGURE 9.5 Schematic of forward-viewing resonant scanning endoscope probe. (Figure reprinted with permission from Liu, X. et al. 2004. *Opt. Lett.* 29 (15): 1763–1765.)

Spiral scanning is obtained by modulating the amplitude of a sine and cosine waveform to each pair of electrodes (Myaing, MacDonald, and Li 2006). The amplitude can be modulated via a sinusoidal waveform. Recently, it has been shown the spiral scanning frequency can be easily tuned by introducing a hybrid cantilever so that the frequency is appropriate with a high-speed swept source for performing real-time 3D endoscopic OCT imaging (Huo et al. 2010).

The same actuator can also produce a Lissajous pattern by adjusting the two-drive waveform frequencies such that the ratio is a rational number. To obtain a pattern with sufficient density the frequencies should be only slightly different. A systematic study of the illumination uniformity of spiral and Lissajous scanning patterns has been conducted, where it was shown that Lissajous scanning patterns provide a more uniform illumination density on the sample (Liang et al. 2012b).

Another method by which *en face* imaging can be performed is via cone beam scanning by using a Risley prism in the distal end optics to offset the beam while using a fiber-optic rotary joint proximally to rotate the catheter and obtain a cone beam image (Fleming et al. 2010). This method has some disadvantages in that the image acquired is a cone shape missing the area inside the cone compared to the aforementioned Lissajous and spiral scanning patterns; however, this catheter proves useful in radio frequency ablation because it can be fabricated without the use of metal components, eliminating a source of interference.

En face imaging probes as seen can often be large in diameter and complex. In an effort to reduce size and provide a simple forward-imaging probe to guide surgical procedures a catheter probe using paired angle rotation scanning (PARS) was implemented (Wu et al. 2006). This catheter provided B-frame type images in depth through the simultaneous rotation of two GRIN lenses in opposite directions; the overall diameter was 1.65 mm but can easily be reduced through the use of smaller GRIN lenses.

9.3.3.4 Needle Probe

Needle imaging probes have been developed to enable interstitial imaging of (soft) solid tissues or organs that were not possible previously with OCT (Li et al. 2000). Figure 9.6 shows two unique designs for a 27 gauge OCT needle probe: Design A (top) is a simple design where a single-mode fiber is polished at 40° and coated to reflect the light beam and allow it to propagate, whereas Design B (bottom) utilized miniature optics at the distal end to provide more control over the imaging beam by adding a GRIN lens and microprism and gluing it together to form one unit (Li et al. 2000).

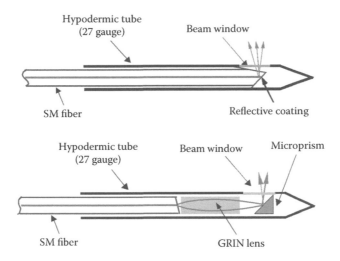

FIGURE 9.6   Needle probe designs. (Figure reprinted with permission from Li, X. et al. 2000. *Opt. Lett.* 25 (20): 1520–1522.)

Other approaches in minimizing the size further came from using a ball lens in Doppler OCT to obviate the use of a microprism and to minimize distal end optics while maintaining cost effectiveness (Yang et al. 2005).

Even more recently Wu et al. have demonstrated a robust needle design based on a similar distal optics as the side-viewing probe, improving on several issues that were previously unaddressed. A small glass tube was used to encase the distal end optics to improve the mechanical and optical robustness of the probe. The glass tube was filled with refractive index matching fluid to minimize the reflection at optical interfaces while the outer surface of the tube was polished flat to minimize the beam distortion caused by the cylindrical lens effect of the glass tube. In addition, the design employed a gold-coated microreflector to deflect the imaging beam which also eased the construction process (Wu et al. 2010).

Recently a 30-gauge, 0.31-mm-diameter needle probe has been fabricated using similar design principles as the larger side-viewing probe discussed previously. The distal end optics consisted of a single-mode fiber and a no-core fiber by which the beam was spread, a GRIN fiber that focused the beam and another piece of no-core fiber which was coated and polished to deflect the beam 90° outside (Lorenser et al. 2011).

9.3.3.5   Multimodal Imaging Probes

As we have seen, OCT produces high resolution images detailing structural information however lacking molecular or biochemical information which is often very useful in disease diagnosis and treatment verification. Naturally, this led to the development of multimodal systems to obtain an additional layer of information. Thus far, OCT has been combined with laser induced fluorescence (LIF) (Barton, Guzman, and Tumlinson 2004; Wall, Bonnema, and Barton 2011), fluorescence spectroscopy (Ryu et al. 2008), single photon excitation fluorescence (Yoo et al. 2011; Mavadia et al 2012; Liang et al. 2012a), and two photon excitation fluorescence (Xi et al. 2012).

One of the challenges in multimodal imaging probe designs is making the catheter compact while still being able to deliver the incident light and collect the returning beams for both OCT and fluorescence imaging, whose wavelengths are typically far from one another.

In the combined OCT and LIF imaging probe, multiple fibers were used to deliver light, one single-mode fiber for OCT and another two multimode fibers with one for fluorescence excitation and one for emission. Each fiber was glued inside a steel tube to which a specially fabricated ball lens was mounted. The imaging beam experienced three total internal reflections within the ball lens and exited perpendicular to the long axis of the endoscope. The overall diameter of the probe was 2 mm with a rigid length of 40 mm. Since the design consisted of multiple fibers, circumferential scanning could not be performed via a traditional fiber-optic rotary joint. Hence, the images were rectangular images obtained by longitudinally translating the probe (Wall, Bonnema, and Barton 2011).

In order to make a more compact probe other groups implemented the multimodal imaging catheters using a double-clad fiber instead of multiple fibers. Double-clad fiber is able to support a single-mode operation for OCT and fluorescence excitation within the core and multimode operation within the inner cladding for more efficient collection of fluorescence emission. One group was able to implement the double-clad fiber for simultaneous OCT and fluorescence spectroscopy through a simple ball lens at the tip of the double-clad fiber and performed imaging *ex vivo* (Ryu et al. 2008). A challenge in working with double-clad fiber is separating signals from the core and inner cladding of the fiber in order to separate the signals obtained from each imaging modality. This can be done in free-space, through the use of a dichroic mirror along with other optical components (Yoo et al. 2011; Xi et al. 2012), or through a popular double-clad fiber coupler (Ryu et al. 2009; Lemire-Renaud et al. 2010; Mavadia et al. 2012; Liang et al. 2012a), which enables the entire system to be fiber optic.

OCT can even be integrated with high-resolution two-photon excitation fluorescence imaging endoscopically. Figure 9.7 depicts a multimodal *en face* endomicroscopy imaging system for simultaneous OCT and two-photon excitation fluorescence imaging. The OCT light and two-photon excitation and emission light can be combined and separated through a wavelength division multiplexer (WDM) after which, the beam is collimated, passed through a dichroic mirror and then coupled into the core of the double-clad fiber. On return, the fluorescence emission, primarily collected by the inner cladding, is reflected by the dichroic mirror to be detected by a PMT, while the OCT return beam travels back through the system and is collected by a balanced detector (Xi et al. 2012).

Similarly, a system was built to perform multimodal OCT and fluorescence imaging through a side-viewing catheter using double-clad fiber (Yoo et al. 2011). This group was able to show that circumferential scanning could be done via a custom made optical rotary joint consisting of free-space components to separate the fluorescence excitation and emission light from the inner cladding of the double-clad fiber. It is important to note that with an optical rotary joint, rotational uniformity is important in both the core and inner cladding of the double-clad fiber and cross-contamination can become a problem if not well aligned. Nevertheless, the use of a

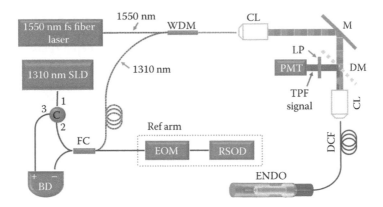

FIGURE 9.7 Multimodal *en face* OCT-two photon endoscope; CL, collimator lens; DM, dichroic mirror; M, mirror; WDM, wave division multiplexer; DCF, double-clad fiber. (Figure reprinted with permission from Xi, J. et al. 2012. *Opt. Lett.* 37 (3): 362–364.)

rotary joint enabled the distal end optics to remain relatively simple and allowed the overall diameter to be no greater than 800 µm.

As mentioned briefly fiber-optic combination and separation of light in double-clad fiber can also be implemented through a double-clad fiber coupler (Ryu et al. 2009; Lemire-Renaud et al. 2010; Mavadia et al. 2012). A double-clad fiber coupler is made to couple light from a fiber (single-mode or double-clad fiber) to the core of the double-clad fiber and on the return path extract the light from the inner cladding into a separate fiber (double-clad or multimode). This approach has several advantages in that the device is not wavelength dependent allowing for the multimodal system to choose fluorescence dyes based on application or novelty.

9.4 EXTENSIONS OF OCT

In this section, we will briefly discuss several extensions of OCT that capture distinct functional information in addition to morphological structure.

9.4.1 POLARIZATION-SENSITIVE OCT

One simplification adopted in our previous description of the OCT theory is that we treated light as a scalar wave rather than a vector field. Yet polarization is another important factor affecting the interference between light backscattered from the tissue and from the reference arm. Polarization-sensitive OCT (PS-OCT) is one extension of OCT that takes into account light polarization. Considering some biological tissues are birefringent, PS-OCT enables assessing of birefringence and optic axis orientation with high spatial resolution. By varying the incident light polarization in a controlled fashion and measuring the backscattered light polarization state, polarization sensitive parameters of tissue can be determined and expressed using Stokes parameters (Everett et al. 1998; de Boer, Milner, and Nelson 1999; Jiao, Yao, and Wang 2000; Saxer et al. 2000; Hitzenberger et al. 2001; Park et al. 2005) or Jones or Mueller matrices (Jiao, Yao, and Wang 2000; Jiao and Wang 2002a, 2002b; Jiao et al. 2003b; Park et al. 2004), and thus diagnostic information can be extracted for various clinical applications, including assessment of burn depth (Park et al. 2001; Jiao et al. 2003a; Srinivas et al. 2004), glaucoma-associated changes in retinal nerve fiber layer (Hee et al. 1992; Cense et al. 2004; Götzinger, Pircher, and Hitzenberger 2005; Naoun et al. 2005), early caries lesion and its progression (Jones, Staninec, and Fried 2004), and articular cartilage for detecting osteoarthritis (Drexler et al. 2001; Li et al. 2005; Patel et al. 2005; Youn et al. 2005).

9.4.2 SPECTROSCOPIC OCT

For simplicity, the local E-field reflection coefficient was assumed to be independent of optical frequency when we derived Equation (9.1). In reality the absorption and scattering properties vary over the spectrum and throughout the tissue, thus the local E-field reflection coefficient will be wavelength dependent. Local spectroscopic features, if extracted, could enhance the contrast of OCT images (Morgner et al. 2000; Yang et al. 2004; Xu et al. 2004; Adler et al. 2004), and provide valuable functional profile of endogenous or exogenous chromophores within the tissue (Faber et al. 2005).

To demonstrate how to extract the spectral information at different depth in the tissue with TD-OCT, the E-field reflectivity $r_S(z)$ in Equation (9.5) should be replaced with $r_S(k, z)$ and we obtain

$$I_D^{AC}(t) \propto r_R \int \int r_S(k,z) S_0(k) \cos(2k(z_R - vt - z)) \, dk \, dz \qquad (9.19)$$

Unlike previous derivation which leads to Equation (9.9), now the power density spectrum $S_0(k)$ is modulated by the wavenumber- or wavelength-dependent reflectivity $r_S(k, z)$. So locally the spectrum of the interference fringe signal is determined by $r_S(k, z) \cdot S_0(k)$. More specifically, when the reference arm scans over a given optical path length z_S, i.e., when $z_R - vt \approx z_S$, the local spectrum is determined by $S_0(k) r_S(k, z_S)$.

To estimate the local spectrum information, the full interference fringe signal rather than only just the envelope must be digitized. Then classical time-frequency joint analysis methods, for example, short-time Fourier transform or wavelet transform, can be employed to extract the local E-field reflection spectrum $r_S(k, z)$ (after normalization by the known power density spectrum of the light source (Morgner et al. 2000)).

The uncertainty principle determines that there always exists a tradeoff between the time (depth) resolution and the spectrum (frequency) resolution. So the optimal time-frequency joint analysis method depends largely on the practical application. And additionally prior knowledge about the spectrum of the light source or the sample can be introduced to conduct model-based analysis (Xu, Kamalabadi, and Boppart 2005). Since most biological components are not spectrally pronounced within the spectrum of commonly used OCT light sources, molecular contrast agents could be considered to further enhance the imaging contrast and molecular specificity (Xu et al. 2004; Yang et al. 2004; Cang et al. 2005; Chen et al. 2005).

9.4.3 DOPPLER OCT

Noninvasive *in vivo* imaging of blood flow is valuable for both biomedical research and clinical diagnosis. Doppler principles can be integrated with OCT, so that the flow information can be retrieved simultaneously with the high-resolution structural imaging, which greatly expands the scope of OCT applications.

Here, we first analyze Doppler detection based on TD-OCT, and let us focus on the PSF caused by a moving particle with reflectivity r_S in the sample arm. Assume at time $t = 0$ the optical path length from the particle to light source is z_S, and it moves with a speed v_S towards the incident probing beam. Mathematically we are setting $r_S(z)$ in Equation (9.1) equal to $r_S \delta(z - (z_S - v_S t))$ for this moving particle. Following the previous derivation, we finally arrive at the PSF as a reduced form of Equation (9.9)

$$PSF_{Doppler}(t) \propto r_R r_S \cdot e^{-16 \ln 2 \cdot (z_R - z_S - v_D t)^2 / \Delta_z^2} \cos(2k_0(z_R - z_S - v_D t)) \qquad (9.20)$$

where $v_D = v - v_S$, and v denotes the constant scanning speed of the mirror in the reference arm. Note that the basic form of this PSF is the same as that in Equation (9.9) except that the carrier frequency changes linearly with the sample velocity v_S. So, again, time-frequency joint analysis can be applied to the full interference fringe signal to retrieve the local carrier frequency of the fringe signal, from which v_D and finally v_S could be deduced.

In general, there exist multiple scatterers moving with different velocities within the coherent volume (equal to lateral imaging spot area multiplied by coherence length). Since the distribution of velocity corresponds to the spectrum of the retrieved local carrier frequency, the centroid of the spectrogram could be used to estimate the average local flow speed. The analysis above constitutes of the basic ideas underlying the spectrogram method used in Doppler OCT (Chen et al. 1997; Izatt et al. 1997; Chen et al. 1999).

The aforementioned limitation of time-frequency joint analysis in Section 9.4.2 again determines the tradeoff between spatial resolution and speed sensitivity. Since the minimum discernible Doppler frequency shift is inversely proportional to the window size Δt used in local frequency analysis, higher speed sensitivity requires larger window size and $v\Delta t$ could dominate the structural axial resolution δ_z given in Equation (9.10), compromising the overall spatial resolution. These limitations inspired the development of the phase-resolved method (Chen et al. 1999; Yang et al. 2002; Ding et al. 2002; Leitgeb, Schmetterer et al. 2003), which analyzes phase delay of the fringe signal between sequential A-line scans to infer the flow speed.

To explain the underlying principles of this phase-resolved method, we assume the A-line scan in Equation (9.20) is repeated after a delay time T. For the second A-line scan, since the starting position of the moving particle in the sample arm has moved by $v_S T$, the fringe signal would be

$$PSF_{Doppler}(t) \propto r_R r_S \cdot e^{-16 \ln 2 \cdot (z_R - (z_S - v_S T) - v_D t)^2 / \Delta_z^2} \cos(2k_0(z_R - (z_S - v_S T) - v_D t)) \qquad (9.21)$$

Comparing Equation (9.21) with Equation (9.20), we see that at the same spatial location, the change in phase (or phase delay) of the carrier wave between two sequential A-line scans is $2k_0 v_S T$. Hence, if the phase delay can be measured, the particle velocity (and flow velocity) v_S can be estimated. In practice, the phase of the carrier wave could be calculated from the complex fringe signal which is the analytic continuation of I^{AC} with the help of Hilbert transform (HT) (Bracewell 1965):

$$A(t)\exp(j\phi(t)) = HT\{I^{AC}(t)\} = I^{AC}(t) - \frac{j}{\pi}\int \frac{I^{AC}(\tau)}{\tau - t} d\tau \qquad (9.22)$$

The power of this phase-resolved method is that it decouples the spatial resolution and velocity sensitivity, thus the imaging speed can be accelerated by more than two orders of magnitude without sacrificing either the spatial resolution or the velocity sensitivity. In Fourier domain OCT, the Fourier transform from k-domain to z-domain will retrieve the complex depth profile of the fringe signal (Leitgeb, Schmetterer et al. 2003; White et al. 2003; Wang et al. 2004; Zhang and Chen 2005). Again we could focus on the PSF caused by a moving particle with reflectivity r_S in the sample arm as when deriving Equation (9.20). This leads to substituting $r_S(z)$ in Equation (9.13) by $r_S \delta(z - (z_S - v_S t))$, and we obtain

$$\hat{I}_D(z) \propto (r_R^2 + r_S^2)\hat{S}_0(z) + r_R r_S \hat{S}_0(z + 2Z) + r_R r_S \hat{S}_0(z - 2Z) \qquad (9.23)$$

where $Z \doteq z_R - (z_S - v_S t)$. With the power density spectrum $\hat{S}_0(z)$ as given in Equation (9.8), the positive half of $\hat{I}_D(z)$ in Equation (9.23) could be expressed as

$$r_R r_S \hat{S}_0(z - 2Z) = r_R r_S e^{-4 \ln 2 \cdot (z - 2Z)^2 / \Delta_z^2} \cdot e^{-jk_0(z - 2Z)} \qquad (9.24)$$

So at a fixed position z, the rate of phase changing is $2k_0 v_S$.

For both TD- and FD-Doppler OCT, there exists a tradeoff between the maximal and minimal detectable flow speed. Recall that the phase change between A-line scans is proportional to $v_S T$. Then given a minimal resolvable phase change, longer delay T enables detection of smaller flow speed v_S, that is, a better sensitivity; on the other hand, since the maximal phase change cannot exceed $\pm\pi$, longer delay T also implies reduced maximal detectable flow speed. In addition, the phase-resolved method suffers from degraded sensitivity to the phase change of moving scatterers in the immediate vicinity of stationary scatterers such as a vessel wall. The vessel size estimated from flow will thus be artificially reduced. To improve the velocity sensitivity and accuracy in determining a vessel size, a moving-scatterer-sensitive method has been developed (Ren and Li 2006).

9.5   OCT APPLICATIONS IN CLINICAL DIAGNOSIS

OCT has many applications in clinical diagnosis and monitoring. The resolution, speed, noninvasiveness, and relatively inexpensive nature as well as the portability of OCT systems make them an ideal option for clinical settings. This technology has traditionally been applied to ophthalmic imaging, but in recent years has been extended to imaging many organs and various clinical conditions. The advent of imaging probes such as endoscopes/catheters and needle probes has enabled OCT imaging to be conveniently integrated with standard medical instruments and protocols for providing additional extra valuable information for diagnosis of diseases, monitoring of therapeutic outcomes, and guidance of biopsy or surgical interventions.

9.5.1 Ophthalmic Imaging

OCT was first used to image the posterior eye structures such as the retina, optic disc, and retinal nerve fiber layer (RNFL) (Huang et al. 1991), and currently physicians have access to commercially available OCT systems in order to image the retina. Recently, OCT has become a popular tool to measure the properties of anterior eye segments such as the cornea, crystalline lens, anterior angle region, and iris (Izatt et al. 1994; Drexler, Hitzenberg et al. 1995a, 1995b; Radhakrishnan et al. 2001). Since then, several commercial OCT systems have become available for a variety of ophthalmic applications.

9.5.2 Cardiovascular Imaging

Cardiovascular imaging poses unique challenges because of the beating of the heart. A rapidly growing field of cardiovascular study is in the study of vulnerable plaques. Vulnerable plaques within the coronary artery are known to be a cause of acute myocardial infarction when they rupture. Traditionally intravascular ultrasound (IVUS) has been used to investigate and characterize the nature of vulnerable plaques. When compared, OCT was found to be able to characterize plaques much better because of its ability to resolve and measure finer structures such as the thin fibrous caps (~65 um) (Jang et al. 2005), as well as distinguish intimal hyperplasia which leads to restenosis (Kawasaki et al. 2006).

Another popular cardiovascular application of OCT is the assessment of stent placement in coronary arteries. Image-guided stent placement, as well as post-stenting surveillance, are necessary to ensure that the stent is in place. OCT has also proven to be useful in this application due to its ability to image long sections of vessels in high resolution and high speed (Guagliumi and Sirbu 2008). Additionally, there is an interest in evaluating tissue structure after stent placement where IVUS is less attractive due to its poor resolution and echoes generated from the stent itself. OCT with a resolution ~10 μm is able to identify tissue prolapse, dissection, and irregular stent struts with much greater detail than IVUS (which has a resolution of ~100 μm) indicating its superiority for evaluation of stent deployment (Bouma et al. 2003).

9.5.3 Gastrointestinal Imaging

Barrett's esophagus (BE) is a condition in which there is a change from stratified squamous to columnar cells in the epithelial lining of the esophagus, usually occurring in patients with chronic reflux. BE appears to be a precancer stage from which

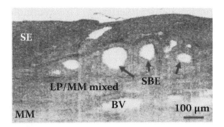 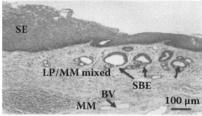

FIGURE 9.8 (Left) Optical coherence tomography image and (right) corresponding histology. Squamous epithelium (SE), subsquamous Barrett's epithelium (SBE), muscularis mucosa (MM), blood vessels (BV), and lamina propria (LP) are all visible. (Figure reprinted with permission from Cobb, M. J. et al. 2010. *Gastrointest. Endosc.* 71 (2): 223–230.)

roughly 17% of patients who exhibit high grade dysplasia at the time of biopsy present with esophageal adenocarcinoma (EAC) (Wang and Sampliner 2008), thusly performing biopsies is becoming increasingly more important in early detection of EAC. The current golden standard is the Seattle protocol consisting of random four quadrant biopsies every 1–2 cm along the length of the esophagus (Sharma et al. 2004; Wang and Sampliner 2008; Manner, Pech, and Ell 2010). There is an increasing need for a technology to assess BE over a large area, aiming to guide biopsy and reduce the associated random sampling error. Furthermore, with the recently growing adoption of preemptive RF ablation treatment of BE (Shaheen et al. 2009), a need for a high-resolution, depth-resolved imaging technology is becoming more critical for assessing subsquamous BE which can be buried beneath normal-looking, newly grown squamous epithelium after RF ablative treatments.

OCT is well positioned to meet the above mentioned technological needs. It has been shown that OCT is able to detect subsquamous BE (shown in Figure 9.8) beneath squamous epithelium in addition to normal tissue structures (such as lamina propria, muscularis mucosa, blood vessel, etc.) (Cobb et al. 2010). With the advent of the balloon imaging catheters, which make it possible to image the esophagus over a large area in a 3D fashion, it has been suggested that OCT can help detect subsquamous BE, EAC, and be used to assess and guide ablative procedures (Caygill et al. 2011).

9.5.4 PULMONARY IMAGING

Most imaging modalities do not have the necessary resolution or contrast to visualize acute changes or early lesions in the lungs. OCT has been compared with conventional histology and shown to be able to distinguish epithelial layer, lamina propria, glandular and cartilaginous structures from human epiglottis (Pitris et al. 1998). Another popular application of OCT is for the study of chronic obstructive pulmonary disorder (COPD). The development of catheters has enabled airway wall thickness to be measured as a way to study airway remodeling and be used for post-treatment monitoring of airway therapy (Coxson et al. 2008). Furthermore, *in vivo* studies with OCT were able to identify malignant airways via thickened mucosa in diseased tissue compared to normal tracheal tissue layers (Hanna et al. 2005). Studies have also been performed to gather OCT images with corresponding histology to form a library in order to determine whether OCT is able to differentiate pathologies, and OCT was found to be able to portray differences in dysplasia, metaplasia, and hyperplasia as well as differentiate invasive cancer from carcinoma *in situ*. Quantitatively, increased epithelial thickness was correlated to severity in histopathology (Lam et al. 2008) as well.

9.5.5 ORAL CAVITY IMAGING

Diseases and conditions in the oral cavity range from dental caries and cavities, to periodontal disease, and to oral cancer. Since diseases have varying symptoms, the use of imaging technologies has become increasingly important for early detection. Conventionally, x-ray has been performed to assess the health of hard tissue (tooth) and along with direct visualization and biopsy to assess lesions in oral mucosa. OCT differentiates between soft and hard tissues as well as identifies periodontal tissue contours, sulcular depth, and connective tissue attachments, which enables physicians to be able to diagnose periodontal disease prior to alveolar bone loss (Otis et al. 2000). Additionally, it has been shown that OCT is able to provide guidance and serve as a diagnostic tool in dental restoration procedures, through its high resolution structural images (Feldchtein et al. 1998). In the case of caries lesions, PS-OCT was able to show correlation between linearly polarized light and the degree of demineralization within dental tissue proving to be immensely useful (Fried et al. 2002; Huysmans, Chew, and Ellwood 2011). In addition to dental applications, OCT has been shown to be effective in identifying lesions in the oral cavity. OCT is able to differentiate stages of squamous cell carcinoma by measuring the epithelial thickness and identifying the boundary between the epithelium and lamina propria in precancerous lesions in human patients (Tsai et al. 2009). Finally, dysplastic lesions also display a loss of stratification in the lower epithelial strata compared to healthy oral mucosa on hamster cheek pouch models (Wilder-Smith et al. 2009).

9.5.6 OTHER APPLICATIONS

In addition to major applications in ophthalmology, cardiovascular, gastrointestinal tract, and pulmonary imaging, there are other applications such as bladder, knee joint and skin imaging with OCT, which we will briefly summarize here.

Bladder imaging for suspected cancer can be performed with catheter/endoscope-based OCT. OCT is able to distinguish the mucosa, lamina propria, and muscularis propria, and differentiate malignant and benign lesions with high accuracy (Lerner et al. 2008). The micro-electro-mechanical system (MEMS)-based imaging probe with a larger beam size made it possible to image large areas. The intrinsic depth-resolved imaging proved to be useful in identifying flat or early benign lesions (Wang et al. 2007).

Knee joint imaging presents a novel application of OCT because it is difficult to image knee joints with traditional imaging modalities such as x-ray and MRI, which do not provide the resolution necessary to image cartilage degeneration, take a long time to image, have radiation dose concern (with x-ray), and potentially are high cost (with MRI). OCT has been shown to be able to distinguish between diseased (osteoarthritic) and normal cartilage tissue during open-knee surgery and in *ex vivo* samples (Li et al. 2005). Furthermore, since cartilage exhibits birefringence, polarization-sensitive OCT (PS-OCT) can be used to monitor changes and detect beginnings of osteoarthritis (Drexler et al. 2001).

Skin is the most easily accessible organ to the naked eye and many imaging devices. OCT represents an attractive imaging modality for skin imaging because it can provide high-resolution images a couple of layers into the skin without tissue removal or matching medium. One application in dermatology is the study of cutaneous wound healing. Studying the process of wound healing is difficult because there is not much technology that allows the process to be examined in high quality without disruption. OCT is able to monitor all of the milestone changes in wound healing such as reepithelialization, formation of the dermal-epidermal junction, dermal thickening, and finally, dermal remodeling (Cobb et al. 2006).

Another important dermatological application of OCT is skin lesion detection in hemangiomas, pemphigus diseases, and epithelial skin cancer. OCT is able to demarcate hemangiomas as signal poor structures under a raised epidermis and calculate the lumen size and detect intraepidermal blisters evident in pemphigus diseases (Konig et al. 2009). Additionally, OCT is able to better quantify and monitor epidermal layer thickness in dermatitis and psoriasis after skin irritation and therapy (Welzel, Bruhns, and Wolff 2003).

9.6 SUMMARY

Optical coherence tomography is a promising and enabling imaging technology that has been applied in many fields already. As a noninvasive technology with a resolution near histology, there are ample benefits of OCT imaging compared to other imaging modalities such as MRI, x-ray/CT, or ultrasound. Additionally, development of innovative imaging probes and light sources has enabled OCT to be used *in vivo* to study many organs/systems from the gastrointestinal (GI) tract, to respiratory tract, and cardiovascular system. OCT can identify lesions in the GI tract, the bladder and lung, via abnormal tissue morphologies compared with healthy tissue. Furthermore, miniature imaging probes have enabled the study of chronic diseases such as COPD by enabling noninvasive monitoring of single airway passages and allowed for the study of vulnerable plaques in the cardiovascular system. The noninvasive nature, high speed, and high-resolution performance of OCT make it a versatile research tool as well as a viable clinical device for diagnosis, treatment, and monitoring.

ACKNOWLEDGMENTS

The authors would like to thank Jiefeng Xi for assistance during the preparation of the material. For other useful discussions over many years, the authors are also grateful to many former and present group members and collaborators, in particular, Dr. Mimi Canto, Dr. Li Huo, Dr. Joo Ha Hwang, Carmen Kut, Dr. Eun Ji Shin, and Dr. Melissa Upton.

REFERENCES

Adler, D., T. Ko, P. R. Herz, and J. G. Fujimoto. 2004. Optical coherence tomography contrast enhancement using spectroscopic analysis with spectral autocorrelation. *Opt. Express* 12 (22): 5487–5501.

Aguirre, A. D., P. Hsiung, T. H. Ko, I. Hartl, and J. G. Fujimoto. 2003. High-resolution optical coherence microscopy for high-speed, *in vivo* cellular imaging. *Opt. Lett.* 28 (21): 2064–2066.

Bachmann, A., R. Leitgeb, and T. Lasser. 2006. Heterodyne Fourier domain optical coherence tomography for full range probing with high axial resolution. *Opt. Express* 14 (4): 1487–1496.

Bajraszewski, T., M. Wojtkowski, M. Szkulmowski, A. Szkulmowska, R. Huber, and A. Kowalczyk. 2008. Improved spectral optical coherence tomography using optical frequency comb. *Opt. Express* 16 (6): 4163–4176.

Barton, J. K., F. Guzman, and A. Tumlinson. 2004. Dual modality instrument for simultaneous optical coherence tomography imaging and fluorescence spectroscopy. *J. Biomed. Opt.* 9 (3): 618–623.

Baumann, B., M. Pircher, E. Götzinger, and C. K. Hitzenberger. 2007. Full range complex spectral domain optical coherence tomography without additional phase shifters. *Opt. Express* 15 (20): 13375–13387.

Beylkin, G. 1995. On the fast Fourier transform of functions with singularities. *Appl. Comput. Harmonic Anal.* 2 (4): 363–381.

Bouma, B. E., and G. J. Tearney. 2002. *Handbook of Optical Coherence Tomography*. New York: Marcel Dekker.

Bouma, B. E., G. J. Tearney, H. Yabushita, M. Shishkov, C. R. Kauffman, D. D. Gauthier, B. D. MacNeill, S. L. Houser, H. T. Aretz, E. F. Halpern, and I. K. Jang. 2003. Evaluation of intracoronary stenting by intravascular optical coherence tomography. *Heart* 89 (3): 317–320.

Bracewell, R. 1965. *The Fourier Transform and Its Applications*. McGraw-Hill.

Brezinski, M. E. 2006. *Optical Coherence Tomography: Principles and Applications*. Amsterdam; Boston: Academic Press.

Cang, H., T. Sun, Z. Y. Li, J. Chen, B. J. Wiley, Y. Xia, and X.D. Li. 2005. Gold nanocages as contrast agents for spectroscopic optical coherence tomography. *Opt. Lett.* 30 (22): 3048–3050.

Caygill, C. P., K. Dvorak, G. Triadafilopoulos, V. N. Felix, J. D. Horwhat, J. H. Hwang, M. P. Upton, X. Li, S. Nandurkar, L. B. Gerson, and G. W. Falk. 2011. Barrett's esophagus: Surveillance and reversal. *Ann. N Y Acad. Sci.* 1232: 196–209.

Cense, B., T. C. Chen, H. B. Park, M. C. Pierce, and J. F. de Boer. 2004. *In vivo* birefringence and thickness measurements of the human retinal nerve fiber layer using polarization-sensitive optical coherence tomography. *J. Biomed. Opt.* 9 (1): 121–125.

Chan, M. C., Y. S. Su, C. F. Lin, and C. K. Sun. 2006. 2.2 microm axial resolution optical coherence tomography based on a 400 nm-bandwidth superluminescent diode. *Scanning* 28 (1): 11–4.

Chang, S. D., E. Murdock, Y. X. Mao, C. Flueraru, and J. Disano. 2011. Stationary-fiber rotary probe with unobstructed 360 degrees view for optical coherence tomography. *Opt. Lett.* 36 (22): 4392–4394.

Chen, J., F. Saeki, B. J. Wiley, H. Cang, M. J. Cobb, Z. Y. Li, L. Au, H. Zhang, M. B. Kimmey, X.D. Li, and Y. Xia. 2005. Gold nanocages: Bioconjugation and their potential use as optical imaging contrast agents. *Nano Lett.* 5 (3): 473–477.

Chen, Z. P., T. E. Milner, S. M. Srinivas, X. Wang, A. Malekafzali, M. J. C. van Gemert, and J. S. Nelson. 1997. Noninvasive imaging of *in vivo* blood flow velocity using optical Doppler tomography. *Opt. Lett.* 22 (14): 1119–1121.

Chen, Z. P., Y. Zhao, S. M. Srinivas, J. S. Nelson, Neal. Prakash, and R. D. Frostig. 1999. Optical Doppler tomography. *IEEE J. Select. Topic. Quantum Electron.* 5 (4): 1134–1142.

Choma, M. A., K. Hsu, and J. A. Izatt. 2005. Swept source optical coherence tomography using an all-fiber 1300-nm ring laser source. *J. Biomed. Opt.* 10 (4): 44009.

Choma, M. A., M. V. Sarunic, C. C. Yang, and J. A. Izatt. 2003. Sensitivity advantage of swept source and Fourier domain optical coherence tomography. *Opt. Express* 11 (18): 2183–2189.

Chudoba, C., J. G. Fujimoto, E. P. Ippen, H. A. Haus, U. Morgner, F. X. Kärtner, V. Scheuer, G. Angelow, and T. Tschudi. 2001. All-solid-state Cr: Forsterite laser generating 14-fs pulses at 1.3 ?m. *Opt. Lett.* 26 (5): 292–294.

Cobb, M. J., Y. Chen, R. A. Underwood, M. L. Usui, J. Olerud, and X. Li. 2006. Noninvasive assessment of cutaneous wound healing using ultrahigh-resolution optical coherence tomography. *J. Biomed. Opt.* 11 (6): 064002.

Cobb, M. J., J. H. Hwang, M. P. Upton, Y. Chen, B. K. Oelschlager, D. E. Wood, M. B. Kimmey, and X. Li. 2010. Imaging of subsquamous Barrett's epithelium with ultrahigh-resolution optical coherence tomography: A histologic correlation study. *Gastrointest. Endosc.* 71 (2): 223–230.

Connelly, M. J. 2002. *Semiconductor Optical Amplifiers.* Boston: Kluwer Academic.

Coxson, H. O., B. Quiney, D. D. Sin, L. Xing, A. M. McWilliams, J. R. Mayo, and S. Lam. 2008. Airway wall thickness assessed using computed tomography and optical coherence tomography. *Am. J. Respir. Crit. Care Med.* 177 (11): 1201–1206.

Davis, A. M., M. A. Choma, and J. A. Izatt. 2005. Heterodyne swept-source optical coherence tomography for complete complex conjugate ambiguity removal. *J. Biomed. Opt.* 10 (6): 064005-6.

de Boer, J. F., B. Cense, B. H. Park, M. C. Pierce, G. J. Tearney, and B. E. Bouma. 2003. Improved signal-to-noise ratio in spectral-domain compared with time-domain optical coherence tomography. *Opt. Lett.* 28 (21): 2067–2069.

de Boer, J. F., T. E. Milner, and J. S. Nelson. 1999. Determination of the depth-resolved Stokes parameters of light backscattered from turbid media by use of polarization-sensitive optical coherence tomography. *Opt. Lett.* 24 (5): 300–302.

Ding, Z. H., Y. Zhao, H. Ren, J. Nelson, and Z. P. Chen. 2002. Real-time phase-resolved optical coherence tomography and optical Doppler tomography. *Opt. Express* 10 (5): 236–245.

Drexler, W., D. Stamper, C. Jesser, X. Li, C. Pitris, K. Saunders, S. Martin, M. B. Lodge, J. G. Fujimoto, and M. E. Brezinski. 2001. Correlation of collagen organization with polarization sensitive imaging of *in vitro* cartilage: Implications for osteoarthritis. *J. Rheumatol.* 28 (6): 1311–1318.

Drexler, W., C. K. Hitzenberg, H. Sattmann, and A. F. Fercher. 1995a. *In vivo* optical coherence tomography and topography of the fundus of the human eye. *Proc. Lasers in Ophthalmology II,* 2330: 134–145.

Drexler, W., C. K. Hitzenberger, H. Sattmann, and A. F. Fercher. 1995b. Measurement of the thickness of fundus layers by partial coherence tomography. *Opt. Eng.* 34 (3): 701–710.

Dutt, A., and V. Rokhlin. 1993. Fast Fourier transforms for nonequispaced data. *SIAM J. Scientific Comput.* 14 (6): 1368–1393.

Eigenwillig, C. M., B. R. Biedermann, G. Palte, and R. Huber. 2008. K-space linear Fourier domain mode locked laser and applications for optical coherence tomography. *Opt. Express* 16 (12): 8916–8937.

Everett, M. J., K. Schoenenberger, B. W. Colston, Jr., and L. B. Da Silva. 1998. Birefringence characterization of biological tissue by use of optical coherence tomography. *Opt. Lett.* 23 (3): 228–230.

Faber, D. J., E. G. Mik, M. C. G. Aalders, and T. G. van Leeuwen. 2005. Toward assessment of blood oxygen saturation by spectroscopic optical coherence tomography. *Opt. Lett.* 30 (9): 1015–1017.

Feldchtein, F. I., G. V. Gelikonov, V. M. Gelikonov, R. R. Iksanov, R. V. Kuranov, A. M. Sergeev, N. D. Gladkova, M. N. Ourutina, J. A. Warren, and D. H. Reitze. 1998. *In vivo* OCT imaging of hard and soft tissue of the oral cavity. *Opt. Express.* 3 (6): 239–250.

Fercher, A. F., K. Mengedoht, and W. Werner. 1988. Eye-length measurement by interferometry with partially coherent light. *Opt. Lett.* 13 (3): 186–188.

Fessler, J. A., and B. P. Sutton. 2003. Nonuniform fast Fourier transforms using min-max interpolation. *IEEE Transactions on Signal Processing* 51 (2): 560–574.

Fleming, C. P., H. Wang, K. J. Quan, and A. M. Rollins. 2010. Real-time monitoring of cardiac radio-frequency ablation lesion formation using an optical coherence tomography forward-imaging catheter. *J. Biomed. Opt.* 15 (3): 030516.

Fried, D., J. Xie, S. Shafi, J. D. Featherstone, T. M. Breunig, and C. Le. 2002. Imaging caries lesions and lesion progression with polarization sensitive optical coherence tomography. *J. Biomed. Opt.* 7 (4): 618–627.

Fu, H. L., Y. X. Leng, M. J. Cobb, K. Hsu, J. H. Hwang, and X. D. Li. 2008. Flexible miniature compound lens design for high-resolution optical coherence tomography balloon imaging catheter. *J. Biomed. Opt.* 13 (6): 060502.

Fujimoto, J. G., and W. Drexler. 2008. Introduction to optical coherence tomography. In *Optical Coherence Tomography: Technology and Applications*, edited by W. Drexler and J. G. Fujimoto: Springer-Verlag, Berlin.

Gelikonov, V., G. Gelikonov, and P. Shilyagin. 2009. Linear-wavenumber spectrometer for high-speed spectral-domain optical coherence tomography. *Optics and Spectroscopy* 106 (3): 459–465.

Gilgen, H. H., R. P. Novak, R. P. Salathe, W. Hodel, and P. Beaud. 1989. Submillimeter optical reflectometry. *Lightwave Technology, Journal of* 7 (8): 1225–1233.

Golubovic, B., B. E. Bouma, G. J. Tearney, and J. G. Fujimoto. 1997. Optical frequency-domain reflectometry using rapid wavelength tuning of a Cr4+: Forsterite laser. *Opt. Lett.* 22 (22): 1704–1706.

Götzinger, E., M. Pircher, and C. K. Hitzenberger. 2005. High speed spectral domain polarization sensitive optical coherence tomography of the human retina. *Opt. Express* 13 (25): 10217–10229.

Guagliumi, G., and V. Sirbu. 2008. Optical coherence tomography: High resolution intravascular imaging to evaluate vascular healing after coronary stenting. *Catheter. Cardiovasc. Interv.* 72 (2): 237–247.

Hanna, N., D. Saltzman, D. Mukai, Z. Chen, S. Sasse, J. Milliken, S. Guo, W. Jung, H. Colt, and M. Brenner. 2005. Two-dimensional and 3-dimensional optical coherence tomographic imaging of the airway, lung, and pleura. *J. Thorac. Cardiovasc. Surg.* 129 (3): 615–622.

Hartl, I., X. D. Li, C. Chudoba, R. K. Ghanta, T. H. Ko, J. G. Fujimoto, J. K. Ranka, and R. S. Windeler. 2001. Ultrahigh-resolution optical coherence tomography using continuum generation in an air-silica microstructure optical fiber. *Opt. Lett.* 26 (9): 608–610.

Hausler, G., and M. W. Lindner. 1998. "Coherence radar" and "spectral radar"—new tools for dermatological diagnosis. *J. Biomed. Opt.* 3 (1): 21–31.

Hee, M. R., D. Huang, E A. Swanson, and J. G. Fujimoto. 1992. Polarization-sensitive low-coherence reflectometer for birefringence characterization and ranging. *J. Opt. Soc. Am. B* 9 (6): 903–908.

Herz, P. R., Y. Chen, A. D. Aguirre, K. Schneider, P. Hsiung, J. G. Fujimoto, K. Madden, J. Schmitt, J. Goodnow, and C. Petersen. 2004. Micromotor endoscope catheter for *in vivo*, ultrahigh-resolution optical coherence tomography. *Opt. Lett.* 29 (19): 2261–2263.

Hitzenberger, C., E. Goetzinger, M. Sticker, M. Pircher, and A. Fercher. 2001. Measurement and imaging of birefringence and optic axis orientation by phase resolved polarization sensitive optical coherence tomography. *Opt. Express* 9 (13): 780–790.

Hu, Z., and A. M. Rollins. 2007. Fourier domain optical coherence tomography with a linear-in-wavenumber spectrometer. *Opt. Lett.* 32 (24): 3525–3527.

Huang, D., E. A. Swanson, C. P. Lin, J. S. Schuman, W. G. Stinson, W. Chang, M. R. Hee, T. Flotte, K. Gregory, C. A. Puliafito, and J. G. Fujimoto. 1991. Optical Coherence Tomography. *Science* 254 (5035): 1178–1181.

Huber, R., M. Wojtkowski, and J. G. Fujimoto. 2006. Fourier Domain Mode Locking (FDML): A new laser operating regime and applications for optical coherence tomography. *Opt. Express* 14 (8): 3225–3237.

Huber, R., M. Wojtkowski, K. Taira, J. Fujimoto, and K. Hsu. 2005. Amplified, frequency swept lasers for frequency domain reflectometry and OCT imaging: Design and scaling principles. *Opt. Express* 13 (9): 3513–3528.

Huo, L., J. F. Xi, Y. C. Wu, and X. D. Li. 2010. Forward-viewing resonant fiber-optic scanning endoscope of appropriate scanning speed for 3D OCT imaging. *Opt. Express* 18 (14): 14375–14384.

Huysmans, M. C., H. P. Chew, and R. P. Ellwood. 2011. Clinical studies of dental erosion and erosive wear. *Caries Res.* 45 (Suppl. 1): 60–68.

Izatt, J. A., M. R. Hee, E. A. Swanson, C. P. Lin, D. Huang, J. S. Schuman, C. A. Puliafito, and J. G. Fujimoto. 1994. Micrometer-scale resolution imaging of the anterior eye *in vivo* with optical coherence tomography. *Arch. Ophthalmol.* 112 (12): 1584–1589.

Izatt, J. A., M. D. Kulkarni, Wang Hsing-Wen, K. Kobayashi, and M. V. Sivak, Jr. 1996. Optical coherence tomography and microscopy in gastrointestinal tissues. *IEEE J. Select. Topic. Quantum Electron.* 2 (4): 1017–1028.

Izatt, J. A., M. D. Kulkarni, S. Yazdanfar, J. K. Barton, and A. J. Welch. 1997. *In vivo* bidirectional color Doppler flow imaging of picoliter blood volumes using optical coherence tomography. *Opt. Lett.* 22 (18): 1439–1441.

Jang, I. K., G. J. Tearney, B. MacNeill, M. Takano, F. Moselewski, N. Iftima, M. Shishkov, S. Houser, H. T. Aretz, E. F. Halpern, and B. E. Bouma. 2005. *In vivo* characterization of coronary atherosclerotic plaque by use of optical coherence tomography. *Circulation* 111 (12): 1551–1555.

Jeon, M., J. Kim, U. Jung, C. K. Lee, W. Jung, and S. A. Boppart. 2011. Full-range k-domain linearization in spectral-domain optical coherence tomography. *Appl. Opt.* 50 (8): 1158–1163.

Jiao, S., and L. V. Wang. 2002a. Jones-matrix imaging of biological tissues with quadruple-channel optical coherence tomography. *J. Biomed. Opt.* 7 (3): 350–358.

Jiao, S., and L. V. Wang. 2002b. Two-dimensional depth-resolved Mueller matrix of biological tissue measured with double-beam polarization-sensitive optical coherence tomography. *Opt. Lett.* 27 (2): 101–103.

Jiao, S., G. Yao, and L. V. Wang. 2000. Depth-resolved two-dimensional stokes vectors of backscattered light and mueller matrices of biological tissue measured with optical coherence tomography. *Appl. Opt.* 39 (34): 6318–6324.

Jiao, S., W. Yu, G. Stoica, and L. V. Wang. 2003a. Contrast mechanisms in polarization-sensitive mueller-matrix optical coherence tomography and application in burn imaging. *Appl. Opt.* 42 (25): 5191–5197.

Jiao, S., W. Yu, G. Stoica, and L. V. Wang. 2003b. Optical-fiber-based Mueller optical coherence tomography. *Opt. Lett.* 28 (14): 1206–1208.

Jones, R. S., M. Staninec, and D. Fried. 2004. Imaging artificial caries under composite sealants and restorations. *J. Biomed. Opt.* 9 (6): 1297–1304.

Kawasaki, M., B. E. Bouma, J. Bressner, S. L. Houser, S. K. Nadkarni, B. D. MacNeill, I. K. Jang, H. Fujiwara, and G. J. Tearney. 2006. Diagnostic accuracy of optical coherence tomography and integrated backscatter intravascular ultrasound images for tissue characterization of human coronary plaques. *J. Am. College Cardiol.* 48 (1): 81–88.

Klein, T., W. Wieser, C. M. Eigenwillig, B. R. Biedermann, and R. Huber. 2011. Megahertz OCT for ultrawide-field retinal imaging with a 1050 nm Fourier domain mode-locked laser. *Opt. Express* 19 (4): 3044–3062.

Konig, K., M. Speicher, R. Buckle, J. Reckfort, G. McKenzie, J. Welzel, M. J. Koehler, P. Elsner, and M. Kaatz. 2009. Clinical optical coherence tomography combined with multiphoton tomography of patients with skin diseases. *J. Biophotonics* 2 (6–7): 389–397.

Lam, S., B. Standish, C. Baldwin, A. McWilliams, J. IeRiche, A. Gazdar, A. I. Vitkin, V. Yang, N. Ikeda, and C. MacAulay. 2008. *In vivo* optical coherence tomography imaging of preinvasive bronchial lesions. *Clin. Cancer Res.* 14 (7): 2006–2011.

Leitgeb, R. A., C. K. Hitzenberger, A. F. Fercher, and T. Bajraszewski. 2003. Phase-shifting algorithm to achieve high-speed long-depth-range probing by frequency-domain optical coherence tomography. *Opt. Lett.* 28 (22): 2201–2203.

Leitgeb, R., C. Hitzenberger, and A. Fercher. 2003. Performance of fourier domain vs. time domain optical coherence tomography. *Opt. Express* 11 (8): 889–894.

Leitgeb, R., L. Schmetterer, W. Drexler, A. Fercher, R. Zawadzki, and T. Bajraszewski. 2003. Real-time assessment of retinal blood flow with ultrafast acquisition by color Doppler Fourier domain optical coherence tomography. *Opt. Express* 11 (23): 3116–3121.

Lemire-Renaud, S., M. Rivard, M. Strupler, D. Morneau, F. Verpillat, X. Daxhelet, N. Godbout, and C. Boudoux. 2010. Double-clad fiber coupler for endoscopy. *Opt. Express* 18 (10): 9755–9764.

Lerner, S. P., A. C. Goh, N. J. Tresser, and S. S. Shen. 2008. Optical coherence tomography as an adjunct to white light cystoscopy for intravesical real-time imaging and staging of bladder cancer. *Urology* 72 (1): 133–137.

Li, X., C. Chudoba, T. Ko, C. Pitris, and J. G. Fujimoto. 2000. Imaging needle for optical coherence tomography. *Opt. Lett.* 25 (20): 1520–1522.

Li, X., S. Martin, C. Pitris, R. Ghanta, D. L. Stamper, M. Harman, J. G. Fujimoto, and M. E. Brezinski. 2005. High-resolution optical coherence tomographic imaging of osteoarthritic cartilage during open knee surgery. *Arthritis Res. Ther.* 7 (2): R318–23.

Liang, S., A. Saidi, J. Jing, G. Liu, J. Li, J. Zhang, C. Sun, J. Narula, and Z. Chen. 2012a. Intravascular atheroxlerotic imaging with combined fluorescence and optical coherence tomography probe based on a double-clad fiber combiner. *J. Biomed. Opt.* 17 (7): 070501.

Liang, W. X., K. Murari, Y. Y. Zhang, Y. P. Chen, M. J. Li, and X. D. Li. 2012b. Increased illumination uniformity and reduced photodamage offered by Lissajous scanning in fiber-optic two-photon endomicroscopy. *J. Biomed. Opt.* 17 (2): 021108.

Liu, X., M. J. Cobb, Y. Chen, M. B. Kimmey, and X. D. Li. 2004. Rapid-scanning forward imaging miniature endoscope for real-time opticalcoherence tomography. *Opt. Lett.* 29 (15): 1763–1765.

Lorenser, D., X. Yang, R. W. Kirk, B. C. Quirk, R. A. McLaughlin, and D. D. Sampson. 2011. Ultrathin side-viewing needle probe for optical coherence tomography. *Opt. Lett.* 36 (19): 3894–3896.

Manner, H., O. Pech, and C. Ell. 2010. Barrett's: evolving techniques for dysplasia detection and endoscopic resection. *Semin. Thorac. Cardiovasc. Surg.* 22 (4): 321–329.

Morgner, U., W. Drexler, F. X. Kartner, X. D. Li, C. Pitris, E. P. Ippen, and J. G. Fujimoto. 2000. Spectroscopic optical coherence tomography. *Opt. Lett.* 25 (2): 111–113.

Murari, K., J. Mavadia, J. F. Xi, and X.D. Li. 2011. Self-starting, self-regulating Fourier domain mode locked fiber laser for OCT imaging. *Biomed. Opt. Express* 2 (7): 2005–2011.

Myaing, M. T., D. J. MacDonald, and X. D. Li. 2006. Fiber-optic scanning two-photon fluorescence endoscope. *Opt. Lett.* 31 (8): 1076–1078.

Naoun, O. K., V. L. Dorr, P. Allé, J. C. Sablon, and A. M. Benoit. 2005. Exploration of the retinal nerve fiber layer thickness by measurement of the linear dichroism. *Appl. Opt.* 44 (33): 7074–7082.

New Scale Technologies, Inc. http: //www.newscaletech.com.

Otis, L. L., M. J. Everett, U. S. Sathyam, and B. W. Colston, Jr. 2000. Optical coherence tomography: A new imaging technology for dentistry. *J. Am. Dent. Assoc.* 131 (4): 511–514.

Pan, Y., H. Xie, and G. K. Fedder. 2001. Endoscopic optical coherence tomography based on a microelectromechanical mirror. *Opt. Lett.* 26 (24): 1966–1968.

Park, B. H., M. C. Pierce, B. Cense, and J. F. de Boer. 2004. Jones matrix analysis for a polarization-sensitive optical coherencetomography system using fiber-optic components. *Opt. Lett.* 29 (21): 2512–2514.

Park, B. H., C. Saxer, S. M. Srinivas, J. S. Nelson, and J. F. de Boer. 2001. *In vivo* burn depth determination by high-speed fiber-based polarization sensitive optical coherence tomography. *J. Biomed. Opt.* 6 (4): 474–479.

Park, B., M. C. Pierce, B. Cense, S. H. Yun, M. Mujat, G. Tearney, B. Bouma, and J. de Boer. 2005. Real-time fiber-based multi-functional spectral-domain optical coherence tomography at 1.3 μm. *Opt. Express* 13 (11): 3931–3944.

Patel, N. A., J. Zoeller, D. L. Stamper, J. G. Fujimoto, and M. E. Brezinski. 2005. Monitoring osteoarthritis in the rat model using optical coherence tomography. *IEEE Trans. Med. Imaging* 24 (2): 155–159.

Pitris, C., M. E. Brezinski, B. E. Bouma, G. J. Tearney, J. F. Southern, and J. G. Fujimoto. 1998. High resolution imaging of the upper respiratory tract with optical coherence tomography—a feasibility study. *Am. J. Respir. Crit. Care Med.* 157 (5): 1640–1644.

Radhakrishnan, S., A. M. Rollins, J. E. Roth, S. Yazdanfar, V. Westphal, D. S. Bardenstein, and J. A. Izatt. 2001. Real-time optical coherence tomography of the anterior segment at 1310 nm. *Arch. Ophthalmol.* 119 (8): 1179–1185.

Ren, H., and X. D. Li. 2006. Clutter rejection filters for optical Doppler tomography. *Opt. Express* 14 (13): 6103–6112.

Ryu, S. Y., H. Y. Choi, J. Na, E. S. Choi, and B. H. Lee. 2008. Combined system of optical coherence tomography and fluorescence spectroscopy based on double-cladding fiber. *Opt. Lett.* 33 (20): 2347–2349.

Ryu, S. Y., H. Y. Choi, J. Na, E. S. Choi, and B. H. Lee. 2009. Simultaneous measurements of optical coherence tomography and fluorescence spectroscopy based on double clad fiber. *Multimodal Biomed. Imaging IV* 7171.

Sarunic, M. V., M. A. Choma, C. C. Yang, and J. A. Izatt. 2005. Instantaneous complex conjugate resolved spectral domain and swept-source OCT using 3x3 fiber couplers. *Opt. Express* 13 (3): 957–967.

Saxer, C. E., J. F. de Boer, B. H. Park, Y. Zhao, Z. P. Chen, and J. S. Nelson. 2000. High-speed fiber based polarization-sensitive optical coherence tomography of *in vivo* human skin. *Opt. Lett.* 25 (18): 1355–1357.

Shaheen, N. J., P. Sharma, B. F. Overholt, H. C. Wolfsen, R. E. Sampliner, K. K. Wang, J. A. Galanko, M. P. Bronner, J. R. Goldblum, A. E. Bennett, B. A. Jobe, G. M. Eisen, M. B. Fennerty, J. G. Hunter, D. E. Fleischer, V. K. Sharma, R. H. Hawes, B. J. Hoffman, R. I. Rothstein, S. R. Gordon, H. Mashimo, K. J. Chang, V. R. Muthusamy, S. A. Edmundowicz, S. J. Spechler, A. A. Siddiqui, R. F. Souza, A. Infantolino, G. W. Falk, M. B. Kimmey, R. D. Madanick, A. Chak, and C. J. Lightdale. 2009. Radiofrequency ablation in Barrett's esophagus with dysplasia. *N. Engl. J. Med.* 360 (22): 2277–2288.

Sharma, P., K. McQuaid, J. Dent, M. B. Fennerty, R. Sampliner, S. Spechler, A. Cameron, D. Corley, G. Falk, J. Goldblum, J. Hunter, J. Jankowski, L. Lundell, B. Reid, N. J. Shaheen, A. Sonnenberg, K. Wang, and W. Weinstein. 2004. A critical review of the diagnosis and management of Barrett's esophagus: The AGA Chicago workshop. *Gastroenterology* 127 (1): 310–330.

Srinivas, S. M., J. de Boer, B. H. Park, K. Keikhanzadeh, H. Huang, J. Zhang, W. Q. Jung, Z. P. Chen, and J. S. Nelson. 2004. Determination of burn depth by polarization-sensitive optical coherence tomography. *J. Biomed. Opt.* 9 (1): 207–212.

Takada, K., I. Yokohama, K. Chida, and J. Noda. 1987. New measurement system for fault location in optical waveguide devices based on an interferometric technique. *Appl. Opt.* 26 (9): 1603–1606.

Targowski, P., I. Gorczynska, M. Szkulmowski, M. Wojtkowski, and A. Kowalczyk. 2005. Improved complex spectral domain OCT for *in vivo* eye imaging. *Opt. Commun.* 249 (1–3): 357–362.

Tearney, G. J., S. A. Boppart, B. E. Bouma, M. E. Brezinski, N. J. Weissman, J. F. Southern, and J. G. Fujimoto. 1996. Scanning single-mode fiber optic catheter-endoscope for optical coherence tomography. *Opt. Lett.* 21 (7): 543–545.

Tran, P. H., D. S. Mukai, M. Brenner, and Z. P. Chen. 2004. *in vivo* endoscopic optical coherence tomography by use of a rotational microelectromechanical system probe. *Opt. Lett.* 29 (11): 1236–1238.

Tsai, M. T., C. K. Lee, H. C. Lee, H. M. Chen, C. P. Chiang, Y. M. Wang, and C. C. Yang. 2009. Differentiating oral lesions in different carcinogenesis stages with optical coherence tomography. *J. Biomed. Opt.* 14 (4): 044028.

Wall, R. A., G. T. Bonnema, and J. K. Barton. 2011. Novel focused OCT-LIF endoscope. *Biomed. Opt. Express* 2 (3): 421–430.

Wang, K. K., and R. E. Sampliner. 2008. Updated guidelines 2008 for the diagnosis, surveillance and therapy of Barrett's esophagus. *Am. J. Gastroenterol.* 103 (3): 788–797.

Wang, L. V., Y. M. Wang, S. Guo, J. Zhang, M. Bachman, G. P. Li, and Z. P. Chen. 2004. Frequency domain phase-resolved optical Doppler and Doppler variance tomography. *Opt. Commun.* 242 (4–6): 345–350.

Wang, R. K. 2007. *In vivo* full range complex Fourier domain optical coherence tomography. *Appl. Phys. Lett.* 90 (5): 054103–3.

Wang, Y. L., M. Bachman, G. P. Li, S. G. Guo, B. J. F. Wong, and Z. P. Chen. 2005. Low-voltage polymer-based scanning cantilever for *in vivo* optical coherence tomography. *Opt. Lett.* 30 (1): 53–55.

Wang, Z. G., C. S. D. Lee, W. C. Waltzer, J. X. Liu, H. K. Xie, Z. J. Yuan, and Y. T. Pan. 2007. *in vivo* bladder imaging with microelectromechanical systems-based endoscopic spectral domain optical coherence tomography. *J. Biomed. Opt.* 12 (3): 034009.

Welzel, J., M. Bruhns, and H. H. Wolff. 2003. Optical coherence tomography in contact dermatitis and psoriasis. *Arch Dermatol. Res.* 295 (2): 50–55.

White, B., M. C. Pierce, N. Nassif, B. Cense, B. Park, G. Tearney, B. Bouma, T. C. Chen, and J. de Boer. 2003. *In vivo* dynamic human retinal blood flow imaging using ultra-high-speed spectral domain optical coherence tomography. *Opt. Express* 11 (25): 3490–3497.

Wilder-Smith, P., K. Lee, S. Guo, J. Zhang, K. Osann, Z. Chen, and D. Messadi. 2009. *In vivo* diagnosis of oral dysplasia and malignancy using optical coherence tomography: Preliminary studies in 50 patients. *Lasers Surg. Med.* 41 (5): 353–357.

Wojtkowski, M., A. Kowalczyk, R. Leitgeb, and A. F. Fercher. 2002. Full range complex spectral optical coherence tomography technique in eye imaging. *Opt. Lett.* 27 (16): 1415–1417.

Wu, J. G., M. Conry, C. H. Gu, F. Wang, Z. Yaqoob, and C. H. Yang. 2006. Paired-angle-rotation scanning optical coherence tomography forward-imaging probe. *Opt. Lett.* 31 (9): 1265–1267.

Wu, Y. C., J. F. Xi, L. Huo, J. Padvorac, E. J. Shin, S. A. Giday, A. M. Lennon, M. I. F. Canto, J. H. Hwang, and X. D. Li. 2010. Robust high-resolution fine OCT needle for side-viewing interstitial tissue imaging. *IEEE J. Select. Topic. Quantum Electron.* 16 (4): 863–869.

Xi, J., Y. Chen, Y. Zhang, K. Murari, M. J. Li, and X. Li. 2012. Integrated multimodal endomicroscopy platform for simultaneous en face optical coherence and two-photon fluorescence imaging. *Opt. Lett.* 37 (3): 362–364.

Xi, J. F., L. Huo, J. Li, and X. D. Li. 2010. Generic real-time uniform K-space sampling method for high-speed swept-Source optical coherence tomography. *Opt. Express* 18 (9): 9511–9517.

Xi, J., L. Huo, Y. Wu, M. J. Cobb, J. H. Hwang, and X. Li. 2009. High-resolution OCT balloon imaging catheter with astigmatism correction. *Opt. Lett.* 34 (13): 1943–1945.

Xu, C., F. Kamalabadi, and S. A. Boppart. 2005. Comparative performance analysis of time-frequency distributions for spectroscopic optical coherence tomography. *Appl. Opt.* 44 (10): 1813–1822.

Xu, C., J. Ye, D. L. Marks, and S. A. Boppart. 2004. Near-infrared dyes as contrast-enhancing agents for spectroscopic opticalcoherence tomography. *Opt. Lett.* 29 (14): 1647–1649.

Yang, C. C., L. E. L. McGuckin, J. D. Simon, M. A. Choma, B. E. Applegate, and J. A. Izatt. 2004. Spectral triangulation molecular contrast optical coherence tomographywith indocyanine green as the contrast agent. *Opt. Lett.* 29 (17): 2016–2018.

Yang, V. X. D., M. L. Gordon, A. Mok, Y. Zhao, Z. P. Chen, R. S. C. Cobbold, B. C. Wilson, and I. Alex Vitkin. 2002. Improved phase-resolved optical Doppler tomography using the Kasai velocity estimator and histogram segmentation. *Opt. Commun.* 208 (4–6): 209–214.

Yang, V. X. D., Y. X. Mao, N. Munce, B. Standish, W. Kucharczyk, N. E. Marcon, B. C. Wilson, and I. A. Vitkin. 2005. Interstitial Doppler optical coherence tomography. *Opt. Lett.* 30 (14): 1791–1793.

Yasuno, Y., S. Makita, T. Endo, G. Aoki, M. Itoh, and T. Yatagai. 2006. Simultaneous B-M-mode scanning method for real-time full-range Fourier domain optical coherence tomography. *Appl. Opt.* 45 (8): 1861–1865.

Yoo, H., J. W. Kim, M. Shishkov, E. Namati, T. Morse, R. Shubochkin, J. R. McCarthy, V. Ntziachristos, B. E. Bouma, F. A. Jaffer, and G. J. Tearney. 2011. Intra-arterial catheter

Youn, J., G. Vargas, B. J. F. Wong, and T. E. Milner. 2005. Depth-resolved phase retardation measurements for laser-assisted non-ablative cartilage reshaping. *Phys. Med. Biol.* 50 (9): 1937.

Youngquist, R. C., S. Carr, and D. E. N. Davies. 1987. Optical coherence-domain reflectometry: A new optical evaluation technique. *Opt. Lett.* 12 (3): 158–160.

Yun, S. H., C. Boudoux, G. J. Tearney, and B. E. Bouma. 2003. High-speed wavelength-swept semiconductor laser with a polygon-scanner-based wavelength filter. *Opt. Lett.* 28 (20): 1981–1983.

Yun, S., G. Tearney, J. de Boer, and B. Bouma. 2004. Removing the depth-degeneracy in optical frequency domain imaging with frequency shifting. *Opt. Express* 12 (20): 4822–4828.

Zara, J. M., S. Yazdanfar, K. D. Rao, J. A. Izatt, and S. W. Smith. 2003. Electrostatic micro-machine scanning mirror for optical coherence tomography. *Opt. Lett.* 28 (8): 628–630.

Zhang, J., and Z. P. Chen. 2005. *In vivo* blood flow imaging by a swept laser source based Fourier domain optical Doppler tomography. *Opt. Express* 13 (19): 7449–7457.

10 Nonlinear Endomicroscopy

Kartikeya Murari and Xingde Li
Johns Hopkins University, Baltimore, Maryland

CONTENTS

10.1   INTRODUCTION

10.1.1   NONLINEAR IMAGING

Nonlinear imaging is an umbrella term that includes imaging techniques reliant on the simultaneous interaction of two or more photons with the sample being observed. Common techniques include two-photon excitation fluorescence (TPEF), coherent anti-stokes Raman spectroscopy (CARS), and second and third harmonic generation (SHG, THG). The beginnings of the field can be traced back to the 1930s when Maria Göepert-Mayer described the principle of multiphoton absorption (Göppert-Mayer 1931). With the creation of the laser in 1960 (Maiman 1960), nonlinear imaging went beyond the realm of theory with the demonstration of SHG and TPEF in 1961 (Franken et al. 1961; Kaiser and Garrett 1961), CARS in 1965 (Maker and Terhune 1965), and THG in 1967 (New and Ward 1967). Over time these techniques have been adapted to microscopy and biomedical imaging (Freund, Deutsch, and Sprecher 1986; Denk, Strickler, and Webb 1990; Barad et al. 1997; Zumbusch, Holtom, and Xie 1999) and are currently some of the most powerful tools (Zipfel, Williams, and Webb 2003; Campagnola and Loew 2003) with a resolution better than the diffraction limited linear imaging techniques.

10.1.2   ADVANTAGES OF NONLINEAR IMAGING

The power of nonlinear imaging stems from the multiphoton interaction requirement. Since photon incidence is a statistical process, one can calculate the number of photons (n) absorbed per incident light pulse in a two-photon process (Denk, Strickler, and Webb 1990), that is,

$$n = \frac{\sigma_{2P} P^2 \pi^2 NA^4}{\tau_p f_p^2 h^2 c^2 \lambda^2}$$

where σ_{2P} is the two-photon absorption cross section, P is the average incident power, NA is the numerical aperture of the imaging lens, τ_p and f_p are the incident light repetition rate and pulsewidth, h is the Planck's constant, c the speed of light, and λ the incident light wavelength. The incident beam intensity drops roughly as the square of the distance from the focal spot. Combined with the quadratic dependence of absorption on incident power, the excitation probability of a fluorophor near the focal spot decreases as the fourth power of the distance of the fluorophor from the focal spot.

The main advantage of this nonlinear dependence on the incident power distance is the potential for high-resolution imaging because signal photons are mainly generated in a tight volume which is smaller than the diffraction limited focal spot. Since all generated photons constitute the signal, regardless of being multiply scattered or not, and therefore can be collected over the full field, a related advantage is that of potentially higher collection efficiency. This is in contrast to confocal imaging where the resolution enhancement is achieved by actively excluding out-of-focus photons by using a pinhole.

While the resolution enhancement occurs in all three dimensions, the most striking benefit can be seen along the optical axis. The high axial resolution is akin to imaging very thin regions of the sample but without actually mechanically sectioning it, and is termed as the ability of optical sectioning. When using fluorescent contrast agents, this gives the additional advantage of confining photobleaching and phototoxicity to the focal spot.

The advantages come with two caveats. Firstly, since the sample is imaged one point at a time, the incident light needs to be scanned across the sample before an image can be reconstructed. Secondly, a very high incident photon flux is needed to counteract the very low probability of multiphoton interaction, leading to the requirement of special light sources.

10.1.3 NONLINEAR ENDOMICROSCOPY

Benchtop implementations of nonlinear imaging, based on traditional laser scanning microscopes, have been around for about two decades. These are used mainly for basic science research and also increasingly for clinical applications, typically for imaging biopsy samples. Their size and complexity is one deterrent for bedside or *in vivo* applications like *in situ* optical biopsy for diagnostics and imaging in freely-moving animals for basic research. To address this, several research groups have been focusing on miniaturizing nonlinear microscopes to make them compatible with standard endoscopy procedures giving birth to the field of nonlinear endomicroscopy (Helmchen et al. 2001; Gobel et al. 2004; Flusberg, Cocker et al. 2005; Flusberg, Lung et al. 2005; Fu et al. 2006; Myaing, MacDonald, and Li 2006; Bao et al. 2008; Engelbrecht et al. 2008; Hoy et al. 2008; Jung et al. 2008; Le Harzic et al. 2008; Wu, Leng et al. 2009; Wu, Xi et al. 2009; Bao, Boussioutas et al. 2010; Bao, Ryu et al. 2010; Wu et al. 2010; Murari et al. 2011; and Fu et al. 2007).

This chapter describes the key aspects of a nonlinear endomicroscope—the light source, light delivery, beam scanning, focusing, light collection, detection, and image reconstruction. Most components have multiple implementations that are compared and contrasted. Finally, we describe representative applications of nonlinear endomicroscopy in TPEF and SHG imaging, followed by a summary and an outlook of the technology.

10.2 KEY ASPECTS

10.2.1 LIGHT SOURCES—ULTRAFAST LASERS

As mentioned earlier, a very high incident photon flux is required to generate the nonlinear signal. Biological samples cannot withstand continuous radiation of this magnitude, thus ultrafast pulsed laser sources are used for nonlinear excitation. By a phenomenon termed mode-locking, these sources typically generate intense pulses of light on the order of a few tens to a few hundreds of femtoseconds at repetition rates nominally in the tens of megahertz with an average power ranging from several hundred milliwatts to watts and a peak power of hundreds of kilowatts. In terms of construction there are two kinds of ultrafast sources.

10.2.1.1 Solid State Lasers

These lasers are characterized by a metal doped crystalline gain medium. They offer a high output power, and depending on the gain medium, are widely tunable in the near infrared (NIR) range. Titanium-doped sapphire lasers are the most common ultrafast lasers used for nonlinear imaging with a 680–1080 nm tuning range. Another example is a chromium-doped forsterite laser with a 1230–1270 nm tuning range.

10.2.1.2 Fiber Lasers

As the name implies, the gain medium for these lasers consists of doped optical fibers. Major advantages over solid state lasers include relatively smaller size, higher flexibility due to fiber coupled output and increased robustness. Femtosecond fiber lasers are generally not inherently tunable and do not have as much output power. Commercial ultrafast fiber lasers are available around 780, 1030, and 1550 nm. Broadband tunability from visible to NIR can be achieved using supercontinuum generation at the expense of reduced power and increased intrinsic noise.

10.2.2 EXCITATION WAVELENGTH RANGES

The most common excitation wavelengths for nonlinear imaging range from about 700 to 900 nm in order to take advantage of the high tissue penetration depth of NIR light with the emitted signal in the visible range (Xu and Webb 1996). This excitation range is well covered by a Ti:Sapphire laser, the most common femtosecond laser. Some groups have used excitation wavelengths longer than 1000 nm for benchtop nonlinear imaging (Chu et al. 2001; McConnell 2007; Yazdanfar et al. 2010). Although water absorption increases for longer wavelength excitation, two factors indicate potentially deeper penetration—(i) reduced scattering for longer wavelengths and (ii) with suitably long excitation, NIR-emitted photons would face lesser attenuation in the collection path. Niche advantages for fiberoptic systems include the ease of dispersion management and the ability to use widely available components developed by the telecommunication industry (Murari et al. 2011). The advantages are offset by a larger diffraction limited focal spot, reducing the resolution, and by the potentially damaging thermal effects owing to the increased water absorption at longer excitation wavelengths.

10.3 BASIC DESIGN CONSIDERATIONS FOR NONLINEAR ENDOMICROSCOPY

A nonlinear endomicroscope system has several parts which will be discussed in this section. Typically these parts include free-space or fiberoptic components for dispersion compensation, single or multiple fibers for light delivery and collection, mechanical scanners for deflecting the beam over the imaging field of view, lenses for focusing the incident light and collecting the nonlinear signal, free-space components to separate the signal from the excitation, and electronics to detect, digitize, and process the signal.

10.3.1 DISPERSION COMPENSATION

Since nonlinear excitation is exponentially related to peak incident power, femtosecond excitation is critical to nonlinear signal generation. Endomicroscopes use optical fiber for light delivery and due to normal dispersion in the fiber, which causes red light to travel faster than blue light, incident pulses get spread out in time. This leads to reduced excitation efficiency making dispersion management critical for nonlinear endomicroscopes. The usual implementation is to precompensate for the dispersion before light is coupled into the fiber. This can be done with a variety of techniques using components that exhibit anomalous dispersion like chirped mirrors, photonic crystal fibers (PCF) (Engelbrecht et al. 2008; Le Harzic et al. 2008), photonic band-gap fiber (PBF) (Flusberg, Lung et al. 2005; Wu et al. 2010; Wu, Leng et al. 2009; Wu, Xi et al. 2009) and prism or grating pairs (Bao, Boussioutas et al. 2010; Bao, Ryu et al. 2010; Bao et al. 2008; Fu et al. 2007; Fu et al. 2006; Gobel et al. 2004; Helmchen et al. 2001; Jung et al. 2008; Myaing, MacDonald, and Li 2006; Treacy 1969; Fork, Martinez, and Gordon 1984). PBF and PCF guide light via a photonic bandgap or closely spaced air holes instead of variations in refractive index. Most of the light is carried in low refractive index regions leading to low nonlinear effects, making these fibers suitable for pulse compression and dispersion compensation. They do not require free-space components, allowing for more robust and compact fiberoptic systems and can also be used for light delivery. However, they operate within a limited range of wavelengths and the amount of compensation depends on their length which cannot be changed in the field. Grating or prism pairs, on the other hand, are easy to tune. Fiberoptic systems around 1.5 μm excitation do not require compensation when utilizing soliton mode transmission and the synergistic effects of anomalous dispersion and self-phase modulation (Murari et al. 2011).

10.3.2 FIBERS

Optical fiber is a crucial part of the endomicroscope, allowing it to be flexible and yet carry light to and from the sample. Light delivery and collection have different requirements. For the excitation, it is crucial to maintain a single mode, so a diffraction limited focal spot can be achieved to ensure high resolution and excitation efficiency, which limits the fiber core size and the NA. On the other hand, for the collection path one needs to maximize signal collection ability requiring large collection areas and a high NA. Two solutions have been proposed to this seeming incompatibility.

10.3.2.1 Multi-Fiber Endomicroscopes

These designs use a single mode fiber (SMF) (Helmchen et al. 2001), a PCF (Engelbrecht et al. 2008; Le Harzic et al. 2008) or a PBF (Flusberg, Lung et al. 2005) for femtosecond light delivery. The latter two allow combining light delivery and dispersion compensation with no free-space requirements. Due to the small core size and low NA, these fibers have poor collection efficiency and are highly sensitive to focal shift between the excitation and emission wavelengths due to chromatic aberrations of the imaging optics (such as the objective lens). Consequently, a large diameter and high NA multimode fiber is used for signal collection. While usage of a large-diameter collection fiber gives high collection efficiency and relaxes lens achromaticity requirements, the use of two fibers complicates the design of the probe head, requiring the integration of dichroic beamsplitters and mirrors, and making it harder to miniaturize.

10.3.2.2 Single-Fiber Endomicroscopes

Approaches using a single fiber for both excitation and collection use a double-clad fiber (DCF) (Bao, Boussioutas et al. 2010; Bao, Ryu et al. 2010; Bao et al. 2008; Murari et al. 2011; Myaing, MacDonald, and Li 2006; Wu et al. 2010; Wu, Leng et al. 2009; Wu, Xi et al. 2009). These fibers have a single mode core, an inner cladding and an outer cladding. The single mode core is similar to the core of a single mode fiber and carries the femtosecond excitation light. The inner cladding is large, on the order of a few hundred microns, and carries the bulk of the nonlinear signal. The large diameter and NA of the inner cladding increases the collection efficiency by 2–3 orders of magnitude compared to a single mode fiber. Figure 10.1 illustrates this advantage, where the large diameter of the inner cladding ($\Phi_{\text{Inner Clad}}$) is comparable to the diameter of the cone of the collected nonlinear signal (Φ_{Signal}) while the core diameter (Φ_{Core}) is much smaller. Thus, a higher proportion of the signal can be collected with a DCF rather than an SMF. In the 700–900 nm range, DCF exhibit normal dispersion and require the use of dispersion management. Although a double-clad photonic crystal fiber (DCPCF) apparently can be used to combine the advantages of a PCF for dispersion compensated delivery and a DCF for collection (Fu et al. 2007, 2006; Jung et al. 2008), the core diameter is generally too large and its numerical aperture is too small, requiring complicated optics to achieve a tight focus which will result in a large probe size.

Figure 10.2 shows the cross section of a double-clad fiber used for a 1.55 μm excitation endomicroscope (Murari et al. 2011). The core, inner cladding, and outer cladding diameters are 8, 180, and 200 μm, respectively. The core diameter is slightly

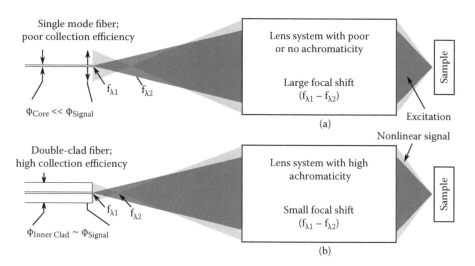

FIGURE 10.1 (See color insert)   Fiber and lens considerations for nonlinear endomicroscopy. (a) Single mode fiber diameter (Φ_{Core}) is much smaller than the diameter of the cone of nonlinear optical signal (Φ_{Signal}) leading to suboptimal collection efficiency. Poor lens achromaticity leads to a large focal shift ($f_{\lambda 1} - f_{\lambda 2}$) between excitation and emission, exacerbating the problem. (b) Large inner cladding diameter of double-clad fiber ($\Phi_{\text{Inner Clad}}$) and lens achromaticity both mitigate the problem, improving collection efficiency.

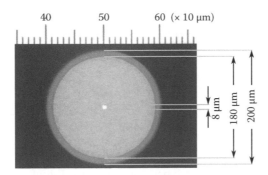

FIGURE 10.2 Photograph of the cross section of a double-clad fiber used for 1.55 μm excitation nonlinear endomicroscopy showing the single mode core and large inner cladding.

smaller than the mode field diameter of a standard 1.55 μm SMF, 10.4 μm (SMF28e, Corning, NY), ensuring single mode excitation light delivery and the large inner cladding increases collection efficiency for the nonlinear signal.

10.3.3 LENSES

The focusing lens is another challenging aspect of nonlinear endomicroscopes. The lens needs to have a high NA and low geometric aberrations for obtaining a near diffraction-limited focal spot size. On the return path, the collection efficiency depends on the NA and also on chromatic aberrations. Chromatic aberration can introduce a loss-inducing focal shift between the excitation and emission wavelengths, which can be up to a few hundred micrometers (Wu, Xi et al. 2009). Figure 10.1 illustrates this difference. A poorly achromatic system, shown in Figure 10.1a, can have a significant focal shift, indicated by $(f_{\lambda 1} - f_{\lambda 2})$, which leads to a larger diameter of the cone of collected signal (Φ_{Signal}) at the endface of the fiber. A better corrected system will exhibit a smaller focal shift and a smaller diameter of the cone of collected signal, as shown in Figure 10.1b. Given equal sized collection fibers, this translates to a significant improvement in the collection efficiency for the achromatic system.

Apart from optical performance, the lens needs to be small enough to be integrated into the probe head. The image NA of the lens (i.e., the NA on the fiber side) should be matched to the NA of the fiber in order to fill the back aperture for the path to minimize the focal spot size and to maximize the coupling efficiency for the collection path. The object NA should be as high as possible to obtain a diffraction limited spot size and high collection efficiency. Other important parameters are the working distance and the imaging medium. Since tissue is not flat and is usually moist or covered with biofluid, an ideal lens should have a long working distance and be designed for usage in water (i.e., a wet lens).

Due to their cylindrical shape and small size, GRIN singlets or compound lenses were the first choice in nonlinear endomicroscope designs (Engelbrecht et al. 2008; Flusberg, Lung et al. 2005; Fu et al. 2007, 2006; Gobel et al. 2004; Jung et al. 2008; Le Harzic et al. 2008; Myaing, MacDonald, and Li 2006; Wu et al. 2010). While they are available in small, submillimeter diameters and are easy to integrate into

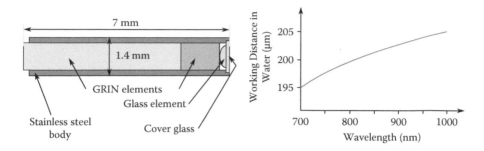

FIGURE 10.3 Mechanical and achromaticity data of a hybrid glass-GRIN compound objective designed for 800 nm excitation nonlinear endomicroscopy. The lens image and object side NA is 0.18 and 0.80, respectively.

the probe head, GRIN lenses, suffer from high chromatic aberrations leading to a low collection efficiency. Miniature compound lenses using multiple glass elements, GRIN lenses or combinations thereof can be designed in order to obtain high NA and a high degree of achromaticity (Bao, Boussioutas et al. 2010; Bao, Ryu et al. 2010; Bao et al. 2008; Helmchen et al. 2001; Murari et al. 2011; Wu, Xi et al. 2009). Glass-GRIN hybrids offer excellent performance, with high NA and very low chromatic shift but are challenging to assemble due to the curved faces of the glass elements, requiring very precise manufacturing.

Figure 10.3 shows mechanical and optical characterization data of an example lens (GT-MO-080-018-810, Grintech GmBH, Germany) designed for usage at 800 nm. The image and object NA are 0.18 and 0.8, respectively. The working distance is 200 μm in water and the lens diameter and length are 1.4 mm and 7 mm, respectively. The lens exhibits superb achromaticity with around 10 μm focal shift across the 700–1000 nm range.

10.3.4 BEAM SCANNERS

Since nonlinear imaging is a point imaging technique, the focused excitation spot must be scanned over the region of interest to form an image. The benchtop approach of galvanometric mirrors cannot be miniaturized enough to be compatible with the probe head leading to the development of several approaches.

10.3.4.1 Piezoelectric Scanners

This technique uses a piezoelectric crystal to generate the mechanical deflection. The light delivery fiber is attached to the crystal leaving a free-standing cantilever, which amplifies the motion and scans the excitation spot. There are two approaches, one using a monolithic cylindrical four-quadrant actuator (Engelbrecht et al. 2008; Murari et al. 2011; Myaing, MacDonald, and Li 2006; Wu et al. 2010; Wu, Leng et al. 2009; Wu, Xi et al. 2009; Liu et al. 2004; Seibel and Smithwick 2002) and another using bending actuators, which are usually bimorphs or trimorphs. The cylindrical actuator can be used to generate spiral or Lissajous scan patterns by driving the orthogonal electrodes pairs with appropriately modulated or frequency and phase-shifted sinusoidal waveforms. A partially cutaway drawing of a tubular scanner is shown in

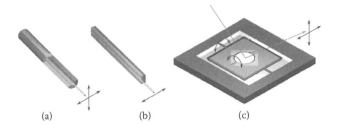

(a) (b) (c)

FIGURE 10.4 Beam scanner designs. (a) Partial cutaway schematic of a monolithic four-quadrant piezoelectric tube with an attached fiber capable of two-dimensional scanning; (b) a piezoelectric bimorph with an attached fiber for one-dimensional scanning and (c) a MEMS mirror with orthogonal gimbal mounts for two-dimensional scanning.

Figure 10.4a along with an attached fiber. A piezoelectric bimorph consists of two dissimilar materials fused to each other and is capable of one-dimensional movement. A schematic is shown in Figure 10.4b. A single bending actuator with some structural modifications may also be used to generate a Lissajous scan (Flusberg, Lung et al. 2005; Helmchen et al. 2001). Another design using two bending actuators arranges them orthogonally and has a fast and slow axis based on the construction (Le Harzic et al. 2008; Rivera, Kobat, and Xu 2011). It can be used to generate a raster scan pattern by driving the fast and slow actuators with sinusoidal and ramp waveforms, respectively. Scanners built using bending actuators offer more flexibility in the scan patterns. The cylindrical scanner, on the other hand, is easier to fabricate and due to its symmetric nature, it is also easier to integrate with the overall probe.

Figure 10.5 shows a photograph of a tubular, four-quadrant monolithic piezoelectric beam scanner with a fiber threaded through the central hole. The overall diameter of the scanner including a housing unit is 2.4 mm. With an exposed fiber length of about 10 mm, the mechanical resonant frequency of the freely-standing cantilever is about 1.3 kHz. With a peak-to-peak drive voltage of about 150 V, the fiber scan range is about 500 µm. Due to the cylindrical symmetry and combining aspects of light delivery and scanning, such scanners can be very compact and, as mentioned before, are easy to integrate into the distal end of the endomicroscope.

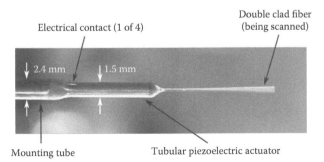

FIGURE 10.5 Photograph of 1.5 mm diameter monolithic, four-quadrant, tubular piezoelectric actuator mounted on a 2.4 mm diameter tube. The double-clad fiber is threaded through the actuator and is being scanned.

10.3.4.2 Micro Electromechanical Systems (MEMS) Scanners

MEMS-based scanners (Fu et al. 2007, 2006; Jung et al. 2008; Hoy et al. 2008; Piyawattanametha et al. 2006; Piyawattanametha and Wang 2010) are built around reflective coatings deposited on micromachined silicon plates with gimbal mounts as shown in Figure 10.4c. These plates act as mirrors which can then be rotated around the mounts using electrostatic forces generated by a comb drive or by electrothermal forces. Additionally, circuits for driving the mirrors can be fabricated on the same silicon die, leading to monolithic scanners. MEMS scanners can implement raster scans very easily and have the potential for region-of-interest windowing but are significantly complicated to fabricate. Also, MEMS scanners tend to be larger than piezoelectric scanners due to the decoupling between light delivery and scanning. Thus, the light beam needs to be folded, which requires some free space for operation. This also makes the design asymmetric and therefore more complicated to integrate into the distal end of an endomicroscope.

10.3.4.3 Scan Patterns

Based on the kind of scanner used, several two-dimensional scan patterns can be achieved for moving the focused excitation spot on the sample. Three common examples are spiral, Lissajous, and raster scans.

The spiral scan, shown in Figure 10.6a, consists of opening and closing spiral trajectories requiring amplitude modulation of the driving waveforms leading to a circular field-of-view. Such scanning is easy to implement on tubular piezoelectric actuators, which are compact and easy to integrate with the probe and image reconstruction is relatively easy. However, it suffers from highly nonuniform spatiotemporal characteristics due to the fact that although angular velocity is maintained, the tangential velocity of the excitation spot increases as it gets further away from the center of rotation. In practice, this leads to higher incident flux and oversampling in the center and lower incident flux and undersampling in the periphery.

The Lissajous scan is equally easy to implement with constant amplitude but frequency shifted drive waveforms leading to a nearly rectangular field-of-view. In

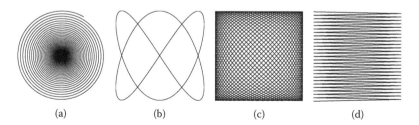

(a)   (b)   (c)   (d)

FIGURE 10.6 Drawings of scan patterns. (a) Spiral pattern showing higher sampling density and fluence in the center. (b) Lissajous pattern with frequency ratio 2:3 showing complex, self-intersecting trajectory. (c) Lissajous pattern with frequency ratio 30:31 showing good sampling density and fluence uniformity. (d) Raster pattern showing excellent uniformity and simple trajectory.

order to obtain a closed curve, the frequency ratio must be rational. Figures 10.6b and 10.6c show Lissajous patterns for drive frequency ratios of 2:3 and 30:31, respectively. The complex self-intersecting trajectory and an improved sampling density and incident fluence can be seen across the field-of-view. A consequence of the complex trajectory is that reconstruction is computationally intensive and is extremely sensitive to drive perturbations.

Scanners to implement raster scanning are relatively harder to fabricate—requiring asymmetric piezoelectric structures or MEMS mirrors, which are also typically larger than tubular piezoelectric actuators and more complicated to integrate with the overall probe design. The drive voltage is often very high (e.g., more than 200 V peak-to-peak). However, the scan pattern, shown in Figure 10.6d, can be easily described in Cartesian space, making image reconstruction very simple. Raster scans have excellent spatiotemporal characteristics with virtually constant incident flux and sampling density within the field-of-view and also the potential for region-of-interest windowing.

10.3.4.4 Z-Scanning

Since nonlinear techniques have the capability for optical sectioning, the incorporation of a mechanism for scanning along the optical axis or Z-scanning enables the construction of a three-dimensional image of the sample. One approach is to use a piston-cylinder based design with either pneumatic or mechanical actuation (Rouse et al. 2004). This requires very high precision manufacturing. A more straightforward way is to use a micromotor (Flusberg, Lung et al. 2005; Helmchen et al. 2001; Le Harzic et al. 2008), which offers low-voltage operation. However, these can be difficult to miniaturize. An alternative is to use an MEMS-based linear actuator or a piezoelectric actuator. These constructs require lever-based mechanical amplification and a few tens to a few hundreds of volts to actuate. Although challenging to fabricate, they offer precise control on the degree of movement. Yet another technique that is easier to implement is using a shape memory alloy (SMA) wire in conjunction with a helical spring (Wu et al. 2010). The SMA wire can be contracted by low-voltage, low-power (~5 mW at 100 mV) electrical heating, compressing the spring. Interrupting the current relaxes the wire and allows the spring to recover, providing bidirectional motion for moving the lens.

10.3.5 Photodetection, Electronics, and Image Reconstruction

As in benchtop systems, photomultiplier tubes (PMT) are used for detecting the nonlinear signal. On the proximal end of the endomicroscope a free-space dichroic mirror is used to separate out the signal from the backscattered and reflected excitation light carried by collection fiber (for multifiber systems) or the DCF (for single fiber systems). The light is slightly focused to match the active photocathode area of the PMT being used. Alternate designs use a fiberoptic coupler for SMF (Bird and Gu 2002), DCPCF (Fu and Gu 2006) and DCF (Bao, Ryu et al. 2010) that avoid the use of a free-space dichroic beamsplitter at the expense of a considerable loss of both the incident and the collected light.

Endomicroscope electronics can be divided into two systems—a driver for the beam scanner and signal detection and conditioning circuits. Both piezoelectric and MEMS-based scanners require a differential voltage drive on the order of several tens to a few hundred volts with modulated or frequency- or phase-shifted sinusoidal or triangular waveforms. Since both scanners present a capacitive load, the power requirements depend on the drive frequency and can be as high as several watts. The high power involved requires careful thermal design of the circuits and proper sealing and electrical insulation of the probe head. On the detection side, the PMT output current is sent to a high-gain (~1 V/μA) and high-bandwidth (~MHz) transimpedance amplifier. The amplifier output is digitized in synchrony with the scanner drive waveforms to allow image reconstruction.

Image reconstruction is based on the scan pattern. While raster scanned data can simply be arranged into a rectangular image, spiral and Lissajous scans require more computation. A polar-to-Cartesian map is required to reconstruct spiral scanned data. Reconstructing Lissajous scan data is the most intensive due to its complex trajectory which requires a precise alignment of the data prior to interpolation.

10.4 MODALITIES AND APPLICATIONS

10.4.1 Two-Photon Excitation Fluorescence (TPEF) Endomicroscopy

TPEF is a fluorescence process where two incident photons simultaneously interact with a fluorophor and transfer the sum of the energy of the two photons to pump the molecule into an excited state. The fluorophor subsequently goes back to the ground state emitting a photon of wavelength longer than half the wavelength of the exciting photons.

Thanks to the wide variety of fluorescent dyes and antibody-antigen chemistry, a very broad range of tissue morphology and physiology can be studied using TPEF. One can also avoid using extrinsic dyes by imaging autofluorescence from biomolecules like NADH and FAD. Since many diseases exhibit cellular abnormalities prior to overt symptoms, TPEF can also be used for diagnostics, which is the key advantage of TPEF endomicroscopy owing to its potential for *in vivo* and *in situ* imaging of internal luminal organs.

10.4.2 Second Harmonic Generation (SHG) Endomicroscopy

SHG is a special case of sum frequency generation where two photons of frequencies f_1 and f_2 interact with noncenterosymmetric structures to generate a photon of wavelength $f_3 = f_1 + f_2$. For SHG, $f_1 = f_2 = f$ and $f_3 = 2f$. In the field of molecular symmetry, a centrosymmetric structure has a point, called the inversion center, relative to which any point (x, y, z) on the structure, there exists an indistinguishable point $(-x, -y, -z)$.

In tissue, noncenterosymmetric structures include structural proteins like collagen, microtubules, and muscle myosin. For imaging these, SHG offers the additional advantage of being a label-free technique that avoids photobleaching and foreign fluorophor-related toxicity. Several diseases and abnormal pathologies cause changes in cytoskeleton and connective tissue, indicating potential usage of SHG imaging for research and diagnostics.

10.4.3 SAMPLE IMAGES

Figure 10.7 shows representative TPEF and SHG images taken with a scanning nonlinear endomicroscope. The endomicroscope featured grating pair dispersion compensation, a single DCF for light delivery and collection, a tubular piezoelectric scanner implementing a spiral scan pattern, and a miniature glass-GRIN hybrid objective lens. The rigid probe head diameter and length was 2.4 mm and 36 mm, respectively. Figure 10.7a shows a schematic of the probe head. Figure 10.7b is a representative TPEF image of acriflavine stained *ex vivo* pig esophagus tissue taken with 810 nm excitation. Figure 10.7c is a representative SHG image of fixed mouse cervical tissue, which was acquired by the same endomicroscope but with 890 nm excitation. Figure 10.7d is a representative image of indocyanine green immunolabeled A431 cells. Again, a similar instrument was used with a different DCF and imaging lens, 1.55 μm excitation, and no dispersion compensation. Figure 10.7e,f shows TPEF and SHG images of the same tissue, but with different fields of view, shown in Figure 10.7b,c, taken using a benchtop nonlinear microscope with a 20 × 0.95 NA objective. These results indicate that in terms of resolution and sensitivity, nonlinear endomicroscopes are approaching their benchtop counterparts while offering tremendous savings in size, weight, flexibility, and cost.

10.5 SUMMARY

In this chapter, we introduced nonlinear optical imaging and related the potential advantages of nonlinear endomicroscopy for clinical and *in situ* translation of the techniques. We reviewed the technologies and instrumentation that form the foundations of the field of nonlinear endomicroscopy. We identified several of the key aspects and components—femtosecond light sources and excitation wavelengths, dispersion compensation for maintaining temporal profiles of the light pulses, optical fibers for light delivery and signal collection, miniature beam scanners for scanning the beam on the sample, miniature lenses for focusing the excitation light and collecting the nonlinear signal, photodetection, signal conditioning and image reconstruction. We discussed all the components and described the various options along with their relative merits. Finally, we touched upon two nonlinear optical techniques—two-photon excitation fluorescence (TPEF) and second harmonic generation (SHG). We showed TPEF and SHG imaging from a single-fiber endomicroscope with a probe diameter of 2.4 mm indicating sensitivity and resolution comparable to a benchtop microscope. We also showed TPEF images from an all-fiberoptic, all-NIR endomicroscope that does not require dispersion compensation.

In closing, the field of nonlinear endomicroscopy is a relatively new but rapidly growing one. It is being driven by advances in mechanical design, MEMS, compact femtosecond fiber lasers, fiberoptics, miniaturized optics, and electronics. It holds great promise to translate the very powerful nonlinear optical imaging techniques into the clinic for diagnostics and also towards basic research in freely-moving animals.

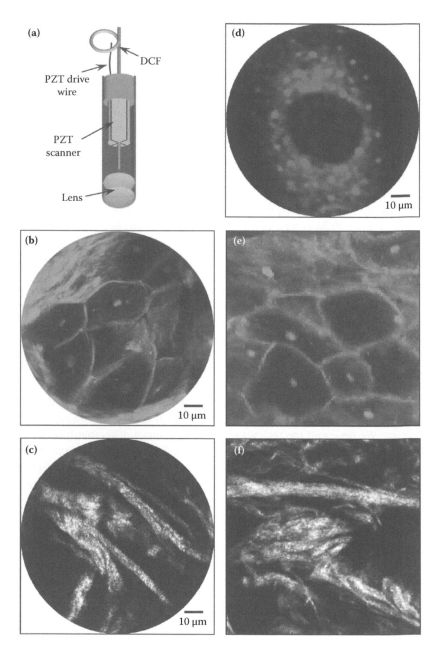

FIGURE 10.7 (a) Schematic of endomicroscope probe head; (b) acriflavine stained *ex vivo* pig esophagus under 810 nm TPEF; (c) fixed mouse cervical tissue under 890 nm SHG; (d) immunolabeled A431 cells under 1.55 μm TPEF; (e,f) benchtop nonlinear microscope images corresponding to (b,c).

ACKNOWLEDGMENTS

The authors would like to thank Dr. Yuying Zhang and Jiefeng Xi for assistance during the preparation of the material. For other useful discussions over many years, the authors are also grateful to many former and present group members and collaborators, in particular, Dr. Yongping Chen, Dr. Ming-Jun Li, Dr. Yicong Wu, Wenxuan Liang, Dr. Kristine Glunde, Dr. Zaver Bhujwalla, Dr. Mala Mahendroo, and Dr. Kate Luby-Phelps.

REFERENCES

Bao, H., A. Boussioutas, R. Jeremy, S. Russell, and M. Gu. 2010. Second harmonic generation imaging via nonlinear endomicroscopy. *Optics Express* 18 (2): 1255–1260.

Bao, H. C., J. Allen, R. Pattie, R. Vance, and M. Gu. 2008. Fast handheld two-photon fluorescence microendoscope with a 475 μm × 475 μm field of view for *in vivo* imaging. *Optics Letters* 33 (12): 1333–1335.

Bao, H., S. Y. Ryu, B. H. Lee, W. Tao, and M. Gu. 2010. Nonlinear endomicroscopy using a double-clad fiber coupler. *Optics Letters* 35 (7): 995–997.

Barad, Y., H. Eisenberg, M. Horowitz, and Y. Silberberg. 1997. Nonlinear scanning laser microscopy by third harmonic generation. *Applied Physics Letters* 70: 922.

Bird, D., and M. Gu. 2002. Compact two-photon fluorescence microscope based on a single-mode fiber coupler. *Optics Letters* 27 (12): 1031–1033.

Campagnola, P. J., and L. M. Loew. 2003. Second-harmonic imaging microscopy for visualizing biomolecular arrays in cells, tissues and organisms. *Nature Biotechnology* 21 (11): 1356–1360.

Chu, S.W., I.H. Chen, T.M. Liu, P.C. Chen, C.K. Sun, and B.L. Lin. 2001. Multimodal nonlinear spectral microscopy based on a femtosecond Cr: Forsterite laser. *Optics Letters* 26 (23): 1909–1911.

Denk, W., J. H. Strickler, and W. W. Webb. 1990. Two-photon laser scanning fluorescence microscopy. *Science* 248 (4951): 73.

Engelbrecht, C. J., R. S. Johnston, E. J. Seibel, and F. Helmchen. 2008. Ultra-compact fiber-optic two-photon microscope for functional fluorescence imaging *in vivo*. *Optics Express* 16 (8): 5556–5564.

Flusberg, B. A., E. D. Cocker, W. Piyawattanametha, J. C. Jung, E. L. M. Cheung, and M. J. Schnitzer. 2005. Fiber-optic fluorescence imaging. *Nature Methods* 2 (12): 941–950.

Flusberg, B. A., J. C. Lung, E. D. Cocker, E. P. Anderson, and M. J. Schnitzer. 2005. *In vivo* brain imaging using a portable 3.9 gram two-photon fluorescence microendoscope. *Optics Letters* 30 (17): 2272–2274.

Fork, R. L., O. E. Martinez, and J. P. Gordon. 1984. Negative dispersion using pairs of prisms. *Optics Letters* 9 (5): 150–152.

Franken, P. A., A. E. Hill, C. W. Peters, and G. Weinreich. 1961. Generation of optical harmonics. *Physical Review Letters* 7 (4): 118–119.

Freund, I., M. Deutsch, and A. Sprecher. 1986. Connective tissue polarity. Optical second-harmonic microscopy, crossed-beam summation, and small-angle scattering in rat-tail tendon. *Biophysical Journal* 50 (4): 693–712.

Fu, L., and M. Gu. 2006. Double-clad photonic crystal fiber coupler for compact nonlinear optical microscopy imaging. *Optics Letters* 31 (10): 1471–1473.

Fu, L., A. Jain, C. Cranfield, H. K. Xie, and M. Gu. 2007. Three-dimensional nonlinear optical endoscopy. *Journal of Biomedical Optics* 12 (4): 040501.

Fu, L., A. Jain, H. K. Xie, C. Cranfield, and M. Gu. 2006. Nonlinear optical endoscopy based on a double-clad photonic crystal fiber and a MEMS mirror. *Optics Express* 14 (3): 1027–1032.

Gobel, W., J. N. D. Kerr, A. Nimmerjahn, and F. Helmchen. 2004. Miniaturized two-photon microscope based on a flexible coherent fiber bundle and a gradient-index lens objective. *Optics Letters* 29 (21): 2521–2523.

Göppert-Mayer, M. 1931. Über elementarakte mit zwei quantensprüngen. *Annalen der Physik* 401 (3): 273–294.

Helmchen, F., M. S. Fee, D. W. Tank, and W. Denk. 2001. A miniature head-mounted two-photon microscope: High-resolution brain imaging in freely moving animals. *Neuron* 31 (6): 903–912.

Hoy, C. L., N. J. Durr, P. Y. Chen, W. Piyawattanametha, H. Ra, O. Solgaard, and A. Ben-Yakar. 2008. Miniaturized probe for femtosecond laser microsurgery and two-photon imaging. *Optics Express* 16 (13): 9996–10005.

Jung, W. Y., S. Tang, D. T. McCormic, T. Q. Xie, Y. C. Ahn, J. P. Su, I. V. Tomov, T. B. Krasieva, B. J. Tromberg, and Z. P. Chen. 2008. Miniaturized probe based on a microelectrome-chanical system mirror for multiphoton microscopy. *Optics Letters* 33 (12): 1324–1326.

Kaiser, W., and C. G. B. Garrett. 1961. Two-photon excitation in CaF_2: Eu^{2+}. *Physical Review Letters* 7 (6): 229–231.

Le Harzic, R., M. Weinigel, I. Riemann, K. Konig, and B. Messerschmidt. 2008. Nonlinear optical endoscope based on a compact two axes piezo scanner and a miniature objective lens. *Optics Express* 16 (25): 20588–20596.

Liu, X. M., M. J. Cobb, Y. C. Chen, M. B. Kimmey, and X. D. Li. 2004. Rapid-scanning forward-imaging miniature endoscope for real-time optical coherence tomography. *Optics Letters* 29 (15): 1763–1765.

Maiman, T. H. 1960. Stimulated optical radiation in ruby masers. *Nature* 187: 493.

Maker, P. D., and R. W. Terhune. 1965. Study of optical effects due to an induced polarization third order in the electric field strength. *Physical Review* 137: 801–818.

McConnell, G. 2007. Nonlinear optical microscopy at wavelengths exceeding 1.4 μm using a synchronously pumped femtosecond-pulsed optical parametric oscillator. *Physics in Medicine and Biology* 52: 717.

Murari, K., Y. Y. Zhang, S. P. Li, Y. P. Chen, M. J. Li, and X.D. Li. 2011. Compensation-free, all-fiber-optic, two-photon endomicroscopy at 1.55 μm. *Optics Letters* 36 (7): 1299–1301.

Myaing, M. T., D. J. MacDonald, and X. D. Li. 2006. Fiber-optic scanning two-photon fluo-rescence endoscope. *Optics Letters* 31 (8): 1076–1078.

New, G. H. C., and J. F. Ward. 1967. Optical third-harmonic generation in gases. *Physical Review Letters* 19 (10): 556–559.

Piyawattanametha, W., R. P. J. Barretto, T. H. Ko, B. A. Flusberg, E. D. Cocker, H. J. Ra, D. S. Lee, O. Solgaard, and M. J. Schnitzer. 2006. Fast-scanning two-photon fluorescence imaging based on a microelectromechanical systems two-dimensional scanning mirror. *Optics Letters* 31 (13): 2018–2020.

Piyawattanametha, W., and T. D. Wang. 2010. MEMS-based dual-axes confocal microendos-copy. *IEEE Journal of Selected Topics in Quantum Electronics,* 16: 804–814.

Rivera, D. R., D. Kobat, and C. Xu. 2011. Miniaturized fiber raster scanner for endoscopy. *Proc. SPIE* 7895: 78950X1-6.

Rouse, A. R., A. Kano, J. A. Udovich, S. M. Kroto, and A. F. Gmitro. 2004. Design and dem-onstration of a miniature catheter for a confocal microendoscope. *Applied Optics* 43 (31): 5763–5771.

Seibel, E. J., and Q. Y. J. Smithwick. 2002. Unique features of optical scanning, single fiber endoscopy. *Lasers in Surgery and Medicine* 30 (3): 177–183.

Treacy, E. 1969. Optical pulse compression with diffraction gratings. *IEEE Journal of Quantum Electronics* 5 (9): 454–458.

Wu, Y. C., Y. X. Leng, J. F. Xi, and X. D. Li. 2009. Scanning all-fiber-optic endomicroscopy system for 3D nonlinear optical imaging of biological tissues. *Optics Express* 17 (10): 7907–7915.

Wu, Y. C., J. F. Xi, M. J. Cobb, and X. D. Li. 2009. Scanning fiber-optic nonlinear endomicroscopy with miniature aspherical compound lens and multimode fiber collector. *Optics Letters* 34 (7): 953–955.

Wu, Y., Y. Zhang, J. Xi, M. J. Li, and X. Li. 2010. Fiber-optic nonlinear endomicroscopy with focus scanning by using shape memory alloy actuation. *Journal of Biomedical Optics* 15 (6): 0506.

Xu, C., and W.W. Webb. 1996. Measurement of two-photon excitation cross sections of molecular fluorophores with data from 690 to 1050 nm. *Journal of the Optical Society of America B* 13 (3): 481–491.

Yazdanfar, S., C. Joo, C. Zhan, M.Y. Berezin, W.J. Akers, and S. Achilefu. 2010. Multiphoton microscopy with near infrared contrast agents. *Journal of Biomedical Optics* 15: 030505.

Zipfel, W. R., R. M. Williams, and W. W. Webb. 2003. Nonlinear magic: Multiphoton microscopy in the biosciences. *Nature Biotechnology* 21 (11): 1369–1377.

Zumbusch, A., G. R. Holtom, and X. S. Xie. 1999. Three-dimensional vibrational imaging by coherent anti-Stokes Raman scattering. *Physical Review Letters* 82 (20): 4142–4145.

11 Micro-Manufacturing Technology of MEMS/MOEMS for Catheter Development

Hadi Mansoor and Mu Chiao
Department of Mechanical Engineering, University of British Columbia, Vancouver, British Columbia, Canada

CONTENTS

11.1 MICRO-ELECTRO-MECHANICAL SYSTEMS

Micro-Electro-Mechanical Systems (MEMS) is a technology that incorporates miniature mechanical components and integrated circuits (IC) to build sensors, actuators, and microsystems (Bustillo et al. 1998). Mass production and cost reduction are possible with MEMS technology compared to traditional sensor/actuator manufacturing technologies. Many MEMS sensors and actuators in micrometer scale have found applications in the automotive industry (Fleming 2001), biotechnology (Bashir 2004), and wireless communications (Rebeiz 2003).

MOEMS or Micro-Opto-Electro-Mechanical Systems are MEMS devices that incorporate optical components such as mirrors (Conant et al. 2000) and lenses (Krogmann et al. 2006). Applications of MOEMS devices include telecommunication (Wu et al. 2006), medical imaging (Mitsui et al. 2006), and consumer products (Urey and Dickensheets 2005). The smallest feature size of a MEMS and MOEMS device usually ranges from a few microns up to a few millimeters.

Conventionally, fabrication of MEMS devices is performed using silicon thin-film technologies similar to IC technology that builds semiconductor circuitry. Batch fabrication of MEMS devices is realized by fabricating several or hundreds of micro devices with similar or different designs on a common silicon wafer in a single fabrication process. This is the same way that IC devices are built and will greatly help in reducing production cost. Because of the small size, MEMS devices can not be fabricated using conventional machining equipment such as cutting, drilling, or forming. Instead, thin-film deposition, photolithography, and chemical etching are the techniques to fabricate MEMS devices.

Deposition is a process of placing thin layers of metals, dielectric materials, semiconductors, or other substances such as photosensitive materials (photoresist) on a substrate. This can be done by physical methods using evaporator, sputtering machine, substrate spinner, or by chemical methods using chemical vapor deposition (CVD) machines. After deposition, thin films are patterned by a process called *photolithography*. It is a process that is used to transfer the MEMS structure design from a photo mask to a photoresist (PR). To make the photo mask, the desired pattern is first generated using computer-aided design (CAD) software. The design is then used to pattern a chromium film deposited on a glass plate by a laser or an electron beam writing system. Next, a layer of PR is spin-coated on the substrate and baked. The thickness of this layer can be controlled by spinning speed and PR viscosity. The photo mask is then placed on the PR, and the PR is selectively exposed to ultraviolet (UV) light. The UV exposure changes the chemical properties of the PR. The substrate is then moved into a next position and UV exposure is carried out again. The process is repeated until all the substrate area is exposed. An illustration of a photolithography process is shown in Figure 11.1. After the UV exposure process, the substrate is placed in a developer solution that etches away either the exposed or unexposed regions of the PR, depending on the type of PR used in the process.

Etching is a process that is frequently used in MEMS fabrication to selectively remove materials that are not protected by PR or any other physical layers above. The materials that can be etched are either the deposited thin films or the substrate itself. Wet etching is performed in a chemical bath. Most wet chemical etchings result in an isotropic etching profile. Wet etching requires simple equipment setup, but precise control on the etching rate is difficult. Dry etching is performed in vacuum chambers using chemicals in gaseous form. Available dry etching methods include reactive ion etching (RIE), deep RIE (DRIE), and plasma etching. Figure 11.2 illustrates basic steps in MEMS fabrication. Further reading of MEMS fabrication processes can be found in other textbooks (Hsu 2008).

11.1.1 Surface Micromachining

The previous section described methods of depositing layers, photo-patterning, and selective etching. However, a unique combination of those techniques is required to realize micro mechanical structures on a substrate. The substrate acts as a mechanical support for the microstructures and can be made of silicon, glass, or plastic.

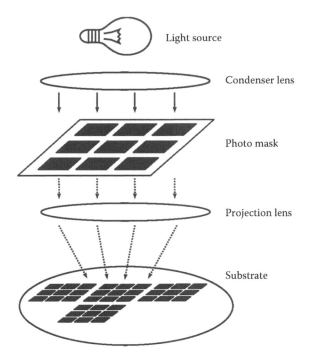

FIGURE 11.1 Illustration of a photolithography process.

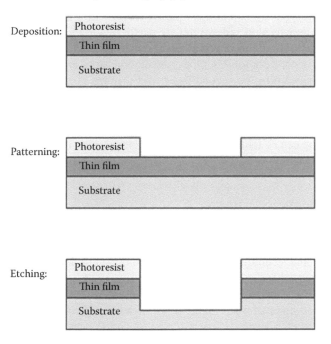

FIGURE 11.2 Basic MEMS fabrication process flow.

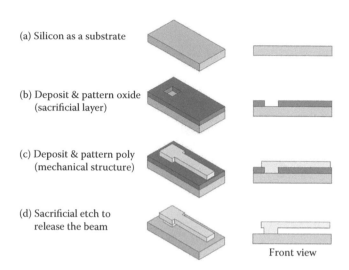

(a) Silicon as a substrate

(b) Deposit & pattern oxide
(sacrificial layer)

(c) Deposit & pattern poly
(mechanical structure)

(d) Sacrificial etch to
release the beam

Front view

FIGURE 11.3   Cantilever beam on silicon substrate.

The micro mechanical structures are the thin films that are deposited and patterned on the surface of the substrate, hence the terminology "surface micromachining." To fabricate free-moving mechanical structures, "sacrificial layers" are needed. Figure 11.3 illustrates an example of the surface micromachining steps required to fabricate a cantilever beam on a silicon substrate.

As shown in Figure 11.3b, a layer of "sacrificial" film is first deposited, and the first photolithography step and etching are performed to form a hole in the film. The second film is deposited that covers the hole and bounds with the substrate to form an "anchor." The second photolithography and etching steps are carried to form a shape of a cantilever beam in the second layer film (Figure 11.3c). Lastly, the sacrificial film is chemically etched, while leaving the cantilever beam intact through etching selectivity (Figure 11.3d). The combinations of material selection, processing chemicals, and temperature in surface micromachining require extensive development. Further reading can be found in "Surface micromachining for microelectromechanical systems" (Bustillo et al. 1998).

11.1.2   BULK MICROMACHINING

In contrast to surface micromachining, "bulk micromachining" directly defines the structures by selective etching of single crystalline silicon substrates. Depending on the chemicals and techniques used, etching can be wet or dry and can have anisotropic or isotropic profiles.

To realize the selective etching, usually a layer of silicon dioxide (SiO_2) is thermally deposited and patterned on a silicon substrate. Then the substrate is exposed to chemicals to etch away the parts of the silicon that are not protected by SiO_2. Figure 11.4 shows schematics of anisotropic and isotropic etching of a silicon substrate along various surface orientations.

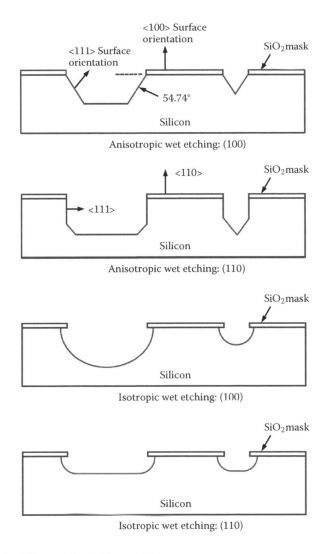

FIGURE 11.4   Silicon wafer etching profiles.

11.1.3   LASER MICROMACHINING

Developed rather more recent than both surface and bulk micromachining technologies, laser micromachining has been used to fabricate microdevices (Miller et al. 2009; Valle et al. 2009). This technique is more flexible in material selection, processing chemicals and temperatures than other micromachining technologies. Furthermore, laser micromachining may be arguably more environmentally friendly since no chemicals are required during the process. A laser with proper optics can be used to cut or drill through thin films. Wavelength, energy, and pulse width of a laser can be properly selected to provide selective machining of a variety of materials, including metals (Hu et al. 2010), polymers (Nguyen and Chan 2006), and ceramics

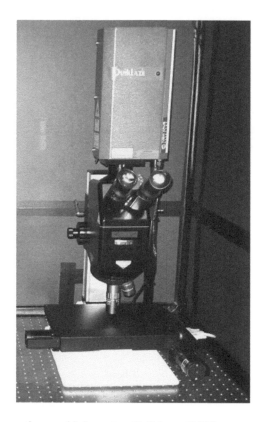

FIGURE 11.5   Laser micromachining setup-QuikLaze 50ST2.

(Samant and Dahotre 2009; Pecholt et al. 2008). In this way a layer of a material can be machined without damaging the layers underneath. In general, "laser writing" is required to fabricate devices a using laser. This can make the process much slower than batch fabrication-type micromachining techniques, such as surface micromachining. This is because the laser removes materials only at a small spot that it is focused at each time. For cutting a line or a shape, usually a CAD file is imported to the laser machining equipment. A motorized stage follows the CAD drawing to achieve the desired shape of cut. Figure 11.5 is photograph of a commercial laser machining system (QuikLaze 50ST2, New Wave Research, USA). Depending on the materials to be cut, the laser wavelength can be selected to be infrared (IR), green, or UV. Figure 11.6 shows a sample cut on a thin film gold using green laser.

11.2   MEMS FOR CATHETER DEVELOPMENT

Various optical microscopy techniques are available for non-destructive imaging of biological tissues. These include optical coherence tomography (OCT) (Huang et al. 1998; Zysk et al. 2007), reflectance-mode confocal microscopy (RCM) (Rajadhyaksha et al. 1995), fluorescence-mode confocal microscopy (FCM) (Laemmel et al. 2004), and multiphoton excitation microscopy (MPM) (Denk et al. 1990; Koenig and

FIGURE 11.6 Green laser used to cut letters UBC in a 200 nm gold evaporated on a silicon substrate. The width of cut is 40 μm.

Riemann 2003). Previous chapters have discussed these techniques. In recent years, researchers have been trying to develop portable microscopic modalities for *in vivo* noninvasive diagnosis of human tissue that could help meet the urgent needs for improving early cancer detection (Kumar et al. 2008). *In vivo* microscopy by all of these modalities has been demonstrated in open field imaging, and in recent years, they have been used in endoscopes for early cancer detection. One of the challenges is to develop miniature catheters that pass through an endoscope's instrument channel, which is typically a few millimeters in diameter. MEMS technology shows a tremendous potential for miniaturization of conventional microscopic modalities. The main advantage of employing MEMS is fabrication of small actuators that enable scanning at the distal tip of the probe. Researchers have been working on developing scanners suitable for endoscopy applications in the past decade (Liu et al. 2007).

In general, there are three types of scanners for optical microscopy; scanning mirrors, lens actuators, and fiber optic scanners. The actuation mechanism of these scanners can be presented in four main categories. This includes electrostatic, electromagnetic, electrothermal, and piezoelectric actuators.

11.2.1 Electrostatic Actuators

Electrostatic actuators are based on the force applied to charged particles by an electric field. One common type of this actuation method is based on two parallel electrodes. In this method one electrode is fixed, and the second one is capable of actuation in an electric field. Petersen developed a scanner (Petersen 1980) by aligning and stacking two micromachined substrates. The upper substrate is made by micromachining single-crystalline silicon to form a center piece that acts as a mirror and two torsional bars to suspend the mirror. The center piece is coated with a layer of aluminum to enhance the reflection. The actual mirror surface area is around 2 × 2 mm². The bottom substrate that can be silicon or glass is etched to form a gap between the mirror and two electrodes that are placed at the bottom of the etched area. The deflection of the mirror is achieved by applying 300–400 volts between the electrodes and the mirror. The demonstrated scanner has a resonance frequency of 15 kHz and is capable of scanning about ±1°.

Another type of electrostatic actuators is based on comb-shape electrodes. Figure 11.7 shows scanning-electron micrograph (SEM) image of a scanning

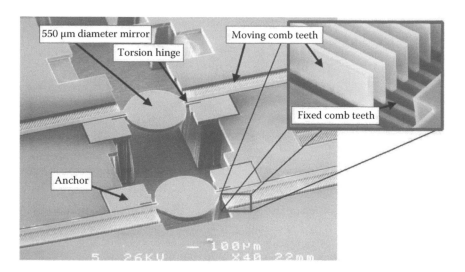

FIGURE 11.7 SEM image of two comb drive micromirrors. The image in the upper right shows a close-up view of the fixed and moving comb teeth. (Reproduced with permission from Conant, R.A. et al. 2000a. *Sens. Actuat.: A. Phys.* 83: 291–296.)

micromirror (Conant et al. 2000) that is actuated by staggered torsional comb actuators. In this design, the lower comb teeth are stationary and fabricated by DRIE of a silicon wafer. The upper comb teeth are fabricated by etching another silicon wafer that is bonded to the first wafer with a thin layer of thermally grown oxide in between. The top wafer also defines the mirror and torsion hinges. Scanning is realized when a voltage is applied between the upper and the lower comb electrodes, exerting a torque on the mirror and causing it to tilt (Figure 11.8). This mirror has a diameter of 550 µm and is capable of scanning ±6.25° at a resonance frequency of 34 kHz.

One disadvantage of staggered comb drives is the requirement for critical alignment of the stationary and moving comb teeth during the manufacturing process. To overcome this issue, angular vertical comb (AVC) actuators are fabricated from a single substrate. Patterson et al., developed a scanning micromirror based on AVC actuators (Patterson et al. 2002). In this approach, the moving comb fingers are attached to the torsion beams that suspend the mirror using photoresist hinges. When a voltage is applied between the movable and stationary fingers, the movable fingers rotate about the torsion beams, hence rotating the mirror. This type of actuator can provide a larger scanning range compared to standard vertical comb drive actuators. The overall size of the scanning micromirror is 1×1 mm². The scanning angle of this micromirror is ±18° with a driving voltage of 21 volts at a resonance frequency of 1.4 kHz.

Kwon demonstrated a 2D transmissive scanner (Kwon and Lee 2002) using microlenses instead of micromirror that offers alignment advantages. They used comb-drive actuators to scan microlenses in X and Y directions. He demonstrated the focus scanning of 75 µm with AC voltage of 10 volts and DC bias of 20 volts.

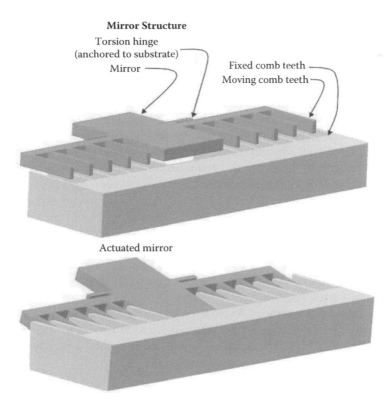

FIGURE 11.8 Schematic of scanning micromirror with torsional comb drives. (Reproduced with permission from Conant, R.A. et al. 2000a. *Sens. Actuat.: A. Phys.* 83: 291–296.)

11.2.2 ELECTROMAGNETIC ACTUATORS

In electromagnetic actuators, a magnetic field is used to move an object. The magnetic field can be from a permanent magnet or an electromagnetic coil. For example, if a current carrying wire is placed in an external magnetic field generated by a permanent magnet or an electromagnet, the wire experiences a force (Lorentz Force) that is the result of interaction between the external magnetic field and the field generated by the current in the wire. Another case of magnetic actuation is when a ferromagnetic material is placed in an external magnetic field. Due to the misalignment of magnetic moments of domains that magnetizes the ferromagnetic material, magnetic force is exerted on the ferromagnetic material. This type of actuation has been used in development of endoscope catheters for various microscopy techniques.

In 1994, a group reported a 2-axis micromirror scanner driven by magnetic fields (Asada et al. 1994). The scanner is micromachined from a silicon substrate. The mirror is connected to a moving frame, which itself is connected to a fixed frame using torsion beams. A pair of planer coils is fabricated around the mirror, and the moving frame and two permanent magnets are placed on both sides of the scanner. Actuation is realized by electromagnetic torque that is produced by the permanent magnets and controlling the electrical current in the planer coils. The scanner has overall dimensions of 7 × 7 mm² and is capable of scanning ±3° and ±1° about X and Y axes at resonant frequencies of 380 and 1450 Hz, respectively.

Mitsui developed another 2-axis magnetically driven micromirror scanner that is suitable for OCT imaging (Mitsui et al. 2006). They used a similar idea as Asada, but a different configuration of magnet and coils to achieve a larger scanning range. In their device, a silicon substrate was etched to form a mirror that is connected to a movable plate by a pair of torsion beams. Similarly, the movable plate is also connected to a fixed frame by the second pair of torsion beams. Two planar coils are fabricated adjacent to the mirror and connected in parallel for X-axis rotation. Four other planar coils for Y-axis rotation are fabricated and connected in parallel on the movable frame. The coil is placed on opposite sides of the mirror and the movable plate spiral in opposite directions. A permanent magnet is also placed under the device. Actuation about X and Y axes is realized by the electromagnetic torques that are produced when controlled electrical current is supplied to the coils. The scanner has the overall size of 7.4 × 9.8 mm² and is capable of ±8° scanning while driven in the static or low frequency (~10 Hz) mode, with a current of ±4.6 mA and ±10.3 mA in the X and Y axes, respectively.

Figure 11.9 shows a catheter developed for high speed OCT imaging (Kim et al. 2007). It incorporates a magnetically actuated micromirror that is capable of 2-axis scanning (Figure 11.10). The actuation is realized by gluing a permanent magnet

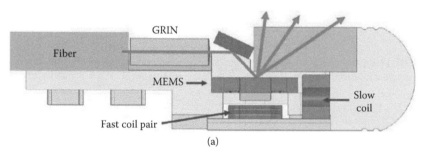

(a)

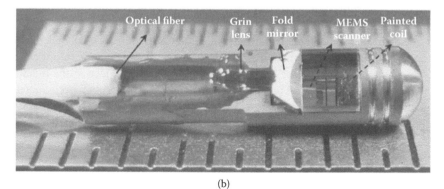

(b)

FIGURE 11.9 Schematic of catheter for OCT. (Reproduced with permission from Kim, K.H. et al. 2007. *Opt. Express.* 15: 18130–18140.)

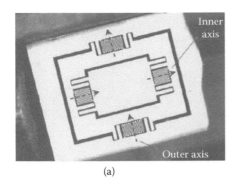
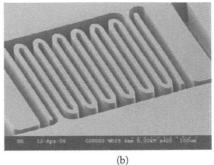

(a)   (b)

FIGURE 11.10 MEMS mirror scanner (a) and a SEM image of a supporting folded flexure (b). (Reproduced with permission from Kim, K.H. et al. 2007. *Opt. Express.* 15: 18130–18140.)

underneath the mirror and applying an external magnetic field using two coils as shown in Figure 11.9a. With a low driving voltage of up to 3 volts, the micromirror scans ± 20° and is capable of operating either statically in both axes or at the resonant frequency (> = 350Hz) for the fast axis. The catheter has a diameter of 2.8 mm and a rigid length of 12 mm.

Another transmissive scanner was developed by mounting a microlens on a nickel platform which is suspended on a moving frame by four V-shaped springs (Siu et al. 2009). The moving frame is also suspended on a fixed frame using three serpentine springs. The microlens is actuated in the out-of-plane direction using an electrostatic force that is generated by applying an AC voltage of 56 volts with a DC bias of 28 volts between the moving and fixed frames. The transverse actuation is realized by applying a 34.5 mT AC magnetic field to the nickel platform. The out-of-plane and transverse scanning ranges are 120 and 163 μm at resonance frequencies of 286 and 480 Hz, respectively.

Out-of-plane actuation of a microlens using electromagnetic fields has also been demonstrated (Mansoor et al. 2011). In this design, a lens is mounted on a microfabricated nickel flexure and is suspended on a frame by folded beams (Figure 11.11a). When an AC magnetic field is applied to the flexure, nickel is magnetized and attracted toward the magnetic field, and the lens is moved in the out-of-plane direction (Figure 11.11b). Figure 11.12 shows two designs of electromagnetic lens actuators. A typical resonant frequency of lens actuator was 544 Hz with a scanning range of 204 μm when a 14 mT AC magnetic field was applied.

11.2.3 ELECTROTHERMAL ACTUATORS

The principle of thermal actuators is based on the linear expansion of materials at elevated temperature. This phenomenon is used in MEMS thermal actuators to fabricate small-size actuation mechanisms. The advantage of electrothermal actuation is mainly in large displacement compared to other types of actuation; however, scanning speed is limited by thermal dissipation rate of the actuator when cooled down from typically a few hundred degrees Celsius. Most demonstrated electrothermal actuators have a scanning speed less than kHz. For out-of-plane actuation, bimorph

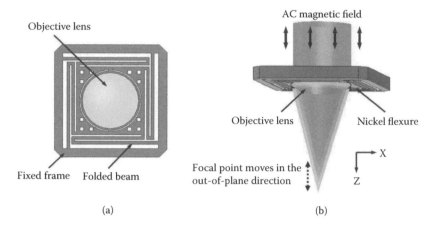

(a)   (b)

FIGURE 11.11 Layout of out-of-plane scanning flexure made of nickel. The diameter of the lens is 1.5 mm and the overall size of the scanner is 3 mm × 3 mm (a) and conceptual illustration of out-of-plane scanning when nickel flexure is placed in an external magnetic field (b). (Reproduce with permission from Mansoor, H. et al. 2011. *Biomed. Microdevices.* 13(4): 641–649.)

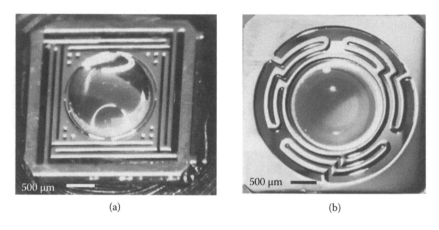

(a)   (b)

FIGURE 11.12 Photograph of two electromagnetic out-of-plane lens actuators fabricated by electroplating technique.

beams consisting of two layers with different coefficients of thermal expansion can be used.

Pan et al. developed a scanning micromirror connected to a frame by a bimorph actuator (Figure 11.13; Pan et al. 2001). The MEMS mirror is 1 × 1 mm² and is fabricated by DRIE of a silicon substrate. The bimorph thermal actuator is composed of a stack of Al and SiO₂ thin films. A layer of polysilicon thermal resistor is embedded in the bimorph mesh and acts as a heat source from joule heating effects. Because of internal residual stress and difference in the thermal expansion of the layers in the hinge, the mirror curls to an angle of about 17° after fabrication. The bimorph

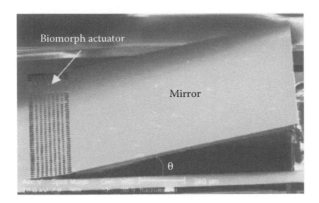

FIGURE 11.13　SEM of the CMOS–MEMS mirror. (Reproduced with permission from Pan, Y. et al. 2001. *Opt. Lett.* 26: 1966–1968.)

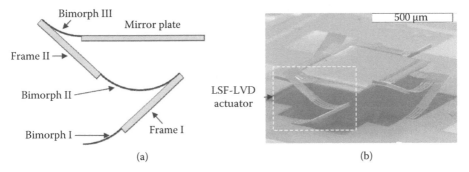

FIGURE 11.14　Cross-sectional view of a thermal bimorph (a), 1D micromirror (b). (Reproduced with permission from Xie, H. et al. 2009. 3D endoscopic optical coherence tomography based on rapid-scanning MEMS mirrors. In *Proc. Communications and Photonics Conference and Exhibition (ACP).* Shanghai, China, 76340X-1.)

actuator can handle voltage up to 33 volts and the mirror has a resonance frequency of 165 Hz and is capable of tilting ±8°.

Another group used a similar approach to develop a 2-axis thermally actuated micromirror (Jain et al. 2004). This mirror is connected to a movable frame that itself is connected to a fixed frame using bimorph cantilever beams. This mirror has dimension of 1×1 mm^2 and can tilt 40° with the applied voltage of 15 volts. The frame can tilt 25° at 17 volts. The resonant frequencies of the mirror and frame actuator structures are 445 and 259 Hz, respectively.

A thermal micromirror scanner for OCT applications has also been developed (Xie et al. 2009) by suspending a micromirror using a series of bimorph cantilever beams (Figure 11.14). The actuation is realized by heating the bimorph beams by directly passing electrical current through them. Figure 11.15 shows the model and photo of a catheter made by the thermal actuator. The diameter of the head is 5.8 mm, and the size of the mirror is 1×1 mm^2. This mirror is capable of scanning ±25° with a driving voltage of less than 10 volts.

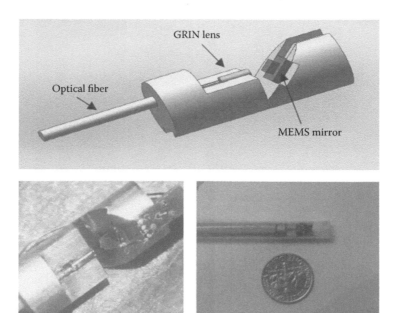

FIGURE 11.15   3D model (a), image of the catheter head (b) and image of the final probe (c) made by the thermal actuator. (Reprinted with permission from Xie, H. et al. 2009. 3D endoscopic optical coherence tomography based on rapid-scanning MEMS mirrors. In *Proc. Communications and Photonics Conference and Exhibition (ACP)*. Shanghai, China, 76340X-1.)

11.2.4   PIEZOELECTRIC ACTUATORS

The piezoelectric effect causes crystalline materials such as quartz and ZnO to expand when charges are applied. The effect is reversible, meaning that a deformation in a piezoelectric material will generate electric charges. The advantage of a piezoelectric actuator is high speed of operation; however, the displacement is relatively small compared to other types of actuation mechanisms. Figure 11.16a shows the schematic of a resonator mirror fabricated by MEMS technology (Goto and Imanaka 1991). A piezoelectric actuator (PZT) is attached to the resonator that is fabricated from a silicon substrate and is capable of scanning an optical beam in two directions shown in Figure 11.16b. For 10 μm PZT actuation, the mirror scans 10° in θ_T and 20° in θ_B directions.

Another piezoelectric actuator was developed by Tsaur et al. (2002). In their design, a micromirror with a dimension of $500 \times 500 \ \mu m^2$ is hinged to four piezoelectric cantilever beams. The overall size of the device is $5 \times 5 \ mm^2$. A scanning angle of 6° at resonance frequency of 1975 Hz was demonstrated.

Transferring available microscopy modalities to endoscopy applications has become a very attractive research area in recent years. Advancement in MEMS/MOEMS technology has allowed significant progress in development of miniature catheters. This has brought a tremendous potential for obtaining high resolution *in vivo* images that can be used to improve the early detection of cancers.

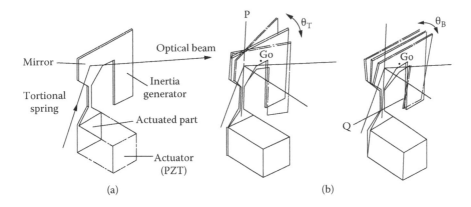

FIGURE 11.16 Schematic of Resonator (a) and deformation modes of resonator (b). (Reproduced with permission from Goto, H. and Imanaka, K. 1991. Super-compact dual-axis optical scanning unit applying a torsional spring resonator driven by a piezoelectric actuator, In *Proc. SPIE.* 1544: 272–281.)

REFERENCES

Asada, N., Matsuki, H., Minami, K., and Esashi, M. 1994. Silicon micromachined two-dimensional galvano optical scanner. *IEEE Trans. Magn.* 30(6): 4647–4649.

Bashir, R. 2004. BioMEMS: State-of-the-art in detection, opportunities and prospects. *Adv. Drug Deliv. Rev.* 56: 1565–1586.

Bustillo, J.M., Howe, R.T., and Muller, R.S. 1998. Surface micromachining for microelectro-mechanical systems. *Proc. IEEE* 86(8): 1552–1574.

Conant, R.A., Hagelin, P.M., Krishnamoorthy, U., Solgaard, O., Lau, K.Y., and Muller, R.S. 2000a. A raster-scanning full-motion video display using polysilicon micromachined mirrors. *Sens. Actuat.: A. Phys.* 83: 291–296.

Conant, R.A., Nee, J.T., Lau, K.Y., and Muller, R.S. 2000b. A flat high-frequency scanning micromirror. In *Proc. Tech. Dig. Solid-State Sensor and Actuator Workshop.* Hilton Head, SC, 6–9.

Denk, W., Strickler, J.H., and Webb, W.W. 1990. Two-photon laser scanning fluorescence microscopy. *Science.* 248: 73.

Fleming, W.J. 2001. Overview of automotive sensors. *IEEE Sens. J.* 1: 296–308.

Goto, H. and Imanaka, K. 1991. Super-compact dual-axis optical scanning unit applying a torsional spring resonator driven by a piezoelectric actuator. In *Proc. SPIE.* 1544: 272–281.

Hsu, T.R. 2008. *MEMS and Microsystems: Design, Manufacture, and Nanoscale Engineering.* Hoboken, NJ: John Wiley & Sons.

Hu, W., Shin, Y.C., and King, G.B. 2010. Micromachining of metals, alloys, and ceramics by picosecond laser ablation. *J. Manuf. Sci. Eng.* 132: 011009.

Huang, D., Swanson, E., Lin, C., Schuman, J., Stinson, W., Chang, W. et al. 1998. Optical coherence tomography. *SPIE Milestone Series MS.* 147: 324–327.

Jain, A., Kopa, A., Pan, Y., Fedder, G.K., and Xie, H. 2004. A two-axis electrothermal micro-mirror for endoscopic optical coherence tomography. *IEEE J. Select. Topics Quantum Electron.* 10: 636–642.

Kim, K.H., Park, B.H., Maguluri, G.N., Lee, T.W., Rogomentich, F.J., Bancu, M.G., Bouma, B.E., de Boer, J.F., and Bernstein, J.J. 2007. Two-axis magnetically-driven MEMS scanning catheter for endoscopic high-speed optical coherence tomography. *Opt. Express.* 15: 18130–18140.

Koenig, K. and Riemann, I. 2003. High-resolution multiphoton tomography of human skin with subcellular spatial resolution and picosecond time resolution. *J. Biomed. Opt.* 8: 432.

Krogmann, F., Mönch, W., and Zappe, H. 2006. A MEMS-based variable micro-lens system. *J. Opt. A: Pure Appl. Opt.* 8: S330–S336.

Kumar, K., Hoshino, K., and Zhang, X. 2008. Handheld subcellular-resolution single-fiber confocal microscope using high-reflectivity two-axis vertical combdrive silicon microscanner. *Biomed. Microdevices.* 10: 653–660.

Kwon, S. and Lee, L. 2002. Stacked two dimensional micro-lens scanner for micro confocal imaging array. In *Proc. IEEE MEMS 2002,* Las Vegas, NV, pp. 483–486.

Laemmel, E., Genet, M., Le Goualher, G., Perchant, A., Le Gargasson, J.F., and Vicaut, E. 2004. Fibered confocal fluorescence microscopy (Cell-viZio™) facilitates extended imaging in the field of microcirculation. *J. Vasc. Res.* 41: 400–411.

Liu, J.T.C., Mandella, M.J., Ra, H., Wong, L.K., Solgaard, O., Kino, G.S., Piyawattanametha, W., Contag, C.H., and Wang, T.D. 2007. Miniature near-infrared dual-axes confocal microscope utilizing a two-dimensional microelectromechanical systems scanner. *Opt. Lett.* 32: 256–258.

Mansoor, H., Zenh, H., and Chaio, M. 2011. Real-time thickness measurement of biological tissues using a microfabricated magnetically-driven lens actuator. *Biomed. Microdevices.* 13(4): 641–649.

Miller, P.R., Aggarwal, R., Doraiswamy, A., Lin, Y.J., Lee, Y.S., and Narayan, R.J. 2009. Laser micromachining for biomedical applications. *JOM J. Miner. Met. Mater. Soc.* 61: 35–40.

Mitsui, T., Takahashi, Y., and Watanabe, Y. 2006. A 2-axis optical scanner driven nonresonantly by electromagnetic force for OCT imaging. *J. Micromech. Microeng.* 16: 2482–2487.

Nguyen, N. and Chan, S.H. 2006. Micromachined polymer electrolyte membrane and direct methanol fuel cells—a review. *J. Micromech. Microeng.* 16: 1.

Pan, Y., Xie, H., and Fedder, G.K. 2001. Endoscopic optical coherence tomography based on a microelectromechanical mirror. *Opt. Lett.* 26: 1966–1968.

Patterson, P.R., Hah, D., Nguyen, H., Toshiyoshi, H., Chao, R., and Wu, M.C. 2002. A scanning micromirror with angular comb drive actuation, In *Proc. IEEE MEMS.* Las Vegas, NV, pp. 544–547.

Pecholt, B., Vendan, M., Dong, Y., and Molian, P. 2008. Ultrafast laser micromachining of 3C-SiC thin films for MEMS device fabrication. *Int. J. Adv. Manuf. Technol.* 39: 239–250.

Petersen, K.E. 1980. Silicon torsional scanning mirror, *IBM J. Res. Dev.* 24: 631–637.

Rajadhyaksha, M., Grossman, M., Esterowitz, D., Webb, R.H., and Anderson, R.R. 1995. *In vivo* confocal scanning laser microscopy of human skin: Melanin provides strong contrast. *J. Invest. Dermatol.* 104: 946–952.

Rebeiz, G.M. 2003. *RF MEMS: Theory, Design, and Technology.* Hoboken, NJ: John Wiley & Sons.

Samant, A.N. and Dahotre, N.B. 2009. Laser machining of structural ceramics—a review. *J. Eur. Ceram. Soc.* 29: 969–993.

Siu, C.P.B., Wang, H., Zeng, H., and Chiao, M., 2009. Dual-axes confocal microlens for Raman spectroscopy. In *Proc. IEEE MEMS.* Sorrento, Italy, pp. 999–1002.

Tsaur, J., Wang, Z.J., Zhang, L., Ichiki, M. Wan, J.W., and Maeda, R. 2002. Preparation and application of lead zirconate titanate (PZT) films deposited by hybrid process: Sol-gel method and laser ablation. *Jpn. J. Appl. Phys.* 41: 6664–6668.

Urey, H. and Dickensheets, D.L. 2005. Display and imaging systems. In *MOEMS: Micro-Opto-Electro-Mechanical Systems. Vol. PM126.* Bellingham: SPIE Press.

Valle, G.D., Osellame, R., and Laporta, P. 2009. Micromachining of photonic devices by femtosecond laser pulses. *J. Opt. A: Pure Appl.Opt.* 11: 013001.

Wu, M.C., Solgaard, O., and Ford, J.E. 2006. Optical MEMS for lightwave communication. *J. Lightwave Technol.* 24: 4433–4454.

Xie, H., Sun, J., Guo, S., and Wu, L. 2009. 3D endoscopic optical coherence tomography based on rapid-scanning MEMS mirrors. In *Proc. Communications and Photonics Conference and Exhibition (ACP)*. Shanghai, China, 76340X-1.

Zysk, A.M., Nguyen, F.T., Oldenburg, A.L., Marks, D.L., and Boppart, S.A. 2007. Optical coherence tomography: A review of clinical development from bench to bedside. *J. Biomed. Opt.* 12: 051403.

Section V

Clinical Applications

12 Clinical Applications in the Lung

Annette McWilliams
British Columbia Cancer Agency Research Centre,
Vancouver, British Columbia, Canada

CONTENTS

12.1 INTRODUCTION

Technological improvements in bronchoscopic optical imaging are expanding our capability to evaluate the airways in a minimally invasive fashion to the microscopic and molecular level.

White light bronchoscopy (WLB) is the most commonly used diagnostic imaging modality for evaluation of the central airways. It uses reflectance imaging that utilizes the optical properties of reflection, backscattering, and absorption of broadband visible light (~400–700 nm). It has been used via flexible bronchoscopy since its development in 1970 by Dr. Shigeto Ikeda and enables rapid macroscopic evaluation of the central airways (Ikeda et al. 1971). Initially, flexible bronchoscopes utilized fiberoptic technology but the subsequent development of color video chip technology (CCD) and incorporation into video bronchoscopes has improved the quality of white light images (Chhajed et al. 2005). Conventional white light bronchoscopy has a high diagnostic yield for proximal tumors but has a limited ability to detect small premalignant and preinvasive bronchial lesions (Chhajed et al. 2005; Woolner et al. 1984).

In addition to optical imaging in the lung, other imaging modalities such as thoracic computed tomography (CT) and endobronchial ultrasound (EBUS) have been developed and are now commonly used in clinical practice. There have been significant advances in CT technology over the last 10 years with the development of multidetector row CT scanners and more recently with quantitative CT scans (Coxson et al. 2009a). Thoracic low dose CT for lung cancer screening has been shown to reduce lung cancer mortality, but it largely detects peripheral adenocarcinoma in the pulmonary parenchyma and is not useful to detect early central lung cancers that are CT occult (The National Lung Screening Trial Research Team 2011; Swensen et al. 2005; McWilliams et al. 2006). In populations where central squamous carcinoma is a significant contributor to lung cancer incidence or in patients with previous aerodigestive tumors or squamous cell lung cancer, evaluation of the central bronchial tree with bronchoscopy remains important (Guardiola et al. 2006; Fujimura et al. 2000; Nakamura et al. 2001; Kennedy et al. 2007; Usuda et al. 2010). Central lung cancer, if found at an early preinvasive stage, has excellent cure rates with local endobronchial therapies (Kennedy et al. 2007). Despite advances in CT technology with improved resolution and reduced radiation dosage, it cannot be currently utilized for evaluation of morphological change within the airway wall, airways beyond the 4th–5th generation or preinvasive lesions in the bronchial tree due to resolution limitations.

EBUS technology, utilizing high frequency ultrasound (20 MHz), has gained rapid clinical application in the lung in the last 5 years. This technology can be applied in a linear fashion via dedicated ultrasound bronchoscopes or in a radial fashion with the use of ultrathin ultrasound probes that can be placed down the working channel of a standard bronchoscope. It can be used to localize peripheral lung lesions and to sample lymph nodes/parabronchial masses for diagnostic and staging purposes (Adams et al. 2009; Steinfort et al. 2011). The layers of the bronchial wall have been well described, and it has an imaging depth of up to 4–5 cm with a spatial resolution of ~0.38 mm (Kurimoto et al. 1999; Kurimoto et al. 2002). Although spatial resolution is limited to the macroscopic level it can be useful to assess the extent of invasion into the bronchial wall of malignant endobronchial lesions (Takahashi et al. 2003).

Current standard white light bronchoscopy and other imaging modalities such as CT and EBUS, although useful, remain limited in the evaluation of pulmonary tissue at higher resolution. There have been a number of exciting optical imaging developments that can be used as an adjunct to WLB to improve the detection and characterization of preinvasive lesions. These techniques can add further biochemical information, evaluation of microvascular networks, and even microscopic imaging of bronchial mucosa and potentially lung parenchyma *in vivo*.

12.2 BRONCHOSCOPIC OPTICAL IMAGING TECHNIQUES

12.2.1 AUTOFLUORESCENCE IMAGING

Autofluorescence bronchoscopy (AFB) was the first of these techniques to be evaluated and clinically established (Lam et al. 1993). The development of AFB in the 1990s led to improved sensitivity to detect preinvasive airway lesions when used as an adjunct to white light imaging. AFB allows rapid examination of large areas of the bronchial mucosa for subtle abnormalities that are not visible under white light examination.

This technique exploits the changes in the autofluorescence of progressively dysplastic bronchial tissue. Endogenous fluorophores in normal bronchial tissues, such as collagen, elastin, NADH and flavins, fluoresce in the green spectrum (480–520 nm) when illuminated by violet/blue light (380–460 nm) (Lam 2006; Wagnieres et al. 2003). As the epithelium changes to more dysplastic pathology there is a reduction in green autofluorescence and proportionately less reduction in red autofluorescence intensity (Lam 2006).

There have been a number of technical improvements in device design since the first device developed by Lam et al. 1993. Most devices now use a filtered lamp for illumination and nonimage intensified sensors (Edell et al. 2009; Häussinger et al. 1999; Zeng et al. 2004; Chiyo et al. 2005). In addition, the original device was used with fiberoptic bronchoscopes, but more recent systems utilize video bronchoscopes with CCD sensors (e.g., SAFE-3000 system, Pentax, Tokyo, Japan; AFI system, Olympus Optical Corp, Tokyo, Japan) (Chiyo et al. 2005; Ikeda et al. 2006). Current commercially available devices use a combination of autofluorescence and reflectance imaging to optimize image quality (Chiyo et al. 2005; Ikeda et al. 2006). Abnormal areas may appear brownish-red, red, purple or magenta while normal areas appear green or light blue depending on the type of reflected light used for the display (Edell et al. 2009; Häussinger et al. 1999; Zeng et al. 2004; Chiyo et al. 2005; Ikeda et al. 2006). (See Figure 12.1.) Some devices allow dual image display

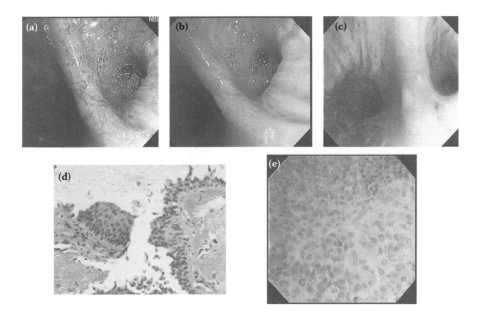

FIGURE 12.1 (See color insert) Evaluation of area of moderate dysplasia at RB6 with various optical imaging modalities. (a) High resolution and narrow band imaging showing complex networks of tortuous vessels, (b) high-resolution white light bronchoscopy, (c) autofluorescence imaging using AFI device; lesion appears magenta on green background, (d) histopathology of biopsy specimen showing moderate dysplasia, (e) endocystoscopic image of lesion. (Courtesy of Dr. Kiyoshi Shibuya, Matsudo City Hospital, Japan.).

of side-by-side white light and autoflourescence images (Ikeda et al. 2006; Lee et al. 2007).

Multiple studies have confirmed the clinical performance of AFB and a recent meta-analysis reported an overall relative sensitivity of 2.04 with combined AFB/WLB compared to WLB alone (Sun et al. 2011). The specificity of combined AFB/WLB was lower than for WLB alone with a pooled specificity of 61% and 80%, respectively (Sun et al. 2011). Quantitative AFB imaging and the use of the red to green fluorescence ratio (R/G) of the target lesion can improve the specificity of AFB to 80% and reduce observer variation (Lee et al. 2009). Combining the R/G ratios with the visual score improves the specificity further to 88% (Lee et al. 2009).

12.2.2 High Magnification White Light Bronchoscopy

The use of high magnification WLB enables the visualization and characterization of mucosal vascular patterns. Changes in the mucosal vasculature and angiogenesis has been described relatively early in lung cancer development (Keith et al. 2000). Shibuya and colleagues first reported this technique in 2002, by using a high magnification bronchovideoscope (XBF 200HM2) to evaluate lesions detected by autofluorescence imaging in 31 high-risk subjects (Shibuya et al. 2002). This bronchoscope, with an outer diameter of 6 mm, combined two systems: a video system for high magnification (fourfold) and a fiberoptic system for orientation of the bronchoscope tip. They were able to visualize the vascular networks in the bronchial mucosa and found that increased vessel growth and complex networks of tortuous vessels were observed in abnormal lesions. They were able to differentiate inflammation from dysplasia (Shibuya et al. 2002).

12.2.3 Narrow Band Imaging

NBI is a technique utilizing selective spectral filtering to white light examination to highlight the vasculature in the bronchial mucosa. (See Figure 12.1.) The two bandwidth ranges of the NBI filter include blue light centered at 415 nm (400–430 nm) and green light at 540 nm (530–550 nm) (Shibuya et al. 2003, 2010; Vincent et al. 2007; Herth et al. 2009). The blue light highlights the superficial capillaries while the green light penetrates more deeply to highlight the larger submucosal blood vessels. The narrow bandwidths reduce the scattering of light from other wavelengths that are present in broad spectrum white light and provide more detailed evaluation of the microvasculature.

Shibuya and colleagues, who first reported description of the vascular patterns in bronchial dysplasia under high magnification WLB, subsequently reported their experience of the combination of high magnification WLB with NBI (Shibuya et al. 2002, 2003). They evaluated 67 lesions detected by autofluorescence imaging in 48 high risk subjects (Shibuya et al. 2003). NBI was useful in the detection of capillary blood vessels in lesions of angiogenic squamous dysplasia at sites of abnormal fluorescence. In 2010, further work by the same group reported the use of combined high resolution bronchoscopy (Olympus BF-6C260) and NBI without the use of autofluorescence localization. They evaluated 37 lesions in 79 subjects with known or suspected lung cancer. Increased vessel growth and complex networks of tortuous vessels was seen in squamous dysplasia, with the addition of dotted vessels in angiogenic squamous dysplasia. Spiral- or screw-type vessels were seen in carcinoma *in situ* (CIS) and microinvasive squamous cell carcinoma. Increasing vessel diameter was found in the progression from dysplasia to carcinoma (Shibuya et al. 2010).

The role of narrow band imaging (NBI) in bronchoscopy is not yet well established. There has been only one published prospective comparative study comparing WLB in combination with either AFB or NBI. Herth and colleagues evaluated 98 lesions detected in 57 high-risk patients without known lung cancer, and found that AFB showed greater sensitivity compared to NBI (65% versus 53%: $p = 0.49$), although this was not significant. AFB had a significantly lower specificity compared to NBI (35% versus 78%; $p < 0.001$) (Herth et al. 2009).

Therefore, NBI may be a useful optical imaging tool without the use of high magnification or high resolution video bronchoscopy. On limited evidence, it appears as sensitive and perhaps more specific than AFB. Further evaluation regarding the performance of NBI with and without autofluorescence imaging, and in direct comparison to newer high definition white light video bronchoscopes is needed to identify its utility.

12.2.4 OPTICAL COHERENCE TOMOGRAPHY

Optical coherence tomography (OCT) is a new and promising imaging method that is similar to ultrasound but utilizes the properties of light rather than sound. It can provide close to histologic resolution of cellular and extracellular structures at and below the tissue surface (Tsuboi et al. 2005; Lam et al. 2008; Ohtani et al. 2012). The axial and lateral resolutions of OCT range from ~5 μm to 30 μm depending on imaging conditions and the imaging depth is 2 to 3 mm. A low-coherence near infrared light is passed into the tissue. Optical interferometry detects the light that is scattered or reflected by the tissue and generates a cross-sectional image along the light direction (Tsuboi et al. 2005; Lam et al. 2008; Ohtani et al. 2012; Coxson and Lam 2009; Coxson et al. 2009b). Two-dimensional images or three-dimensional volumetric images can be created by scanning the light beam over the tissue (Ohtani et al. 2012). centimeters of airway can be imaged in a few seconds and helical scans can be performed allowing a cylindrical volume of tissue to be imaged (Ohtani et al. 2012). This combination of resolution and imaging depth is ideal for examining pre-neoplastic changes originating in epithelial tissues or tumors involving smaller airways.

The imaging procedure is performed *in vivo* using miniature fiberoptic probes that can be inserted down the instrument channel during standard bronchoscopic examination under conscious sedation. Tissue contact is not required to obtain the images. (See Figure 12.2.) OCT is a safe technique with no associated risks from the near-infrared light source. Compared to ultrasound it does not require a liquid coupling medium, and thus is more compatible with airway imaging.

Tsuboi et al. described the initial evaluation of OCT in the lung in seven *ex vivo* lobectomy specimens and *in vivo* examination of five patients (Tsuboi et al. 2005). OCT imaging revealed a layered bronchial wall structure in a normal bronchus that

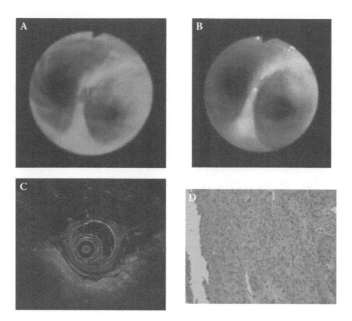

FIGURE 12.2 (See color insert)   Area of carcinoma *in situ* at RB3 in 76 yo male with COPD. (A) Area of CIS under autofluorescence imaging with OncoLife system (Xillix, Richmond, BC, Canada), (B) Area of CIS with standard white light bronchoscopy, (C) OCT image of area of CIS showing thickened epithelium with intact basement membrane, (D) Histopathology showing carcinoma *in situ*. (Courtesy of Professor Stephen Lam and Dr. Keishi Ohtani, BC Cancer Agency, Vancouver, Canada.).

was lost in lung cancer. In the peripheral lung, air-containing alveoli are imaged by OCT as a honeycomb structure beyond the bronchial wall. (See Figure 12.3.)

The ability of OCT to differentiate invasive cancer versus CIS or dysplasia was investigated by Lam et al. in 148 subjects undergoing bronchoscopy (Lam et al. 2008). The thickness of the epithelial layer increases as the epithelium changes from normal/hyperplasia to metaplasia to dysplasia/CIS. These changes were quantitatively measured by OCT and demonstrated that the invasive carcinoma was significantly

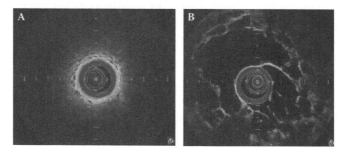

FIGURE 12.3   OCT of peripheral airway. (A) Peripheral OCT in smoker without COPD, (B) peripheral OCT in smoker with severe COPD showing significant emphysema and destruction of alveoli.

thicker than the carcinoma *in situ* ($p = 0.004$) and dysplasia was significantly thicker than metaplasia or hyperplasia ($p = 0.002$). The nuclei became more readily visible in high-grade dysplasia or CIS. The basement membrane was still intact in CIS but became discontinuous or no longer visible with invasive cancer. Different grades of dysplasia could not be distinguished using the OCT technology in the study.

OCT has also been evaluated for the assessment of small airway changes in COPD (Coxson et al. 2008; Coxson and Lam 2009). Coxson et al. compared measurements of airway wall dimensions in 44 smokers with both OCT and CT in specific bronchial segments. A strong correlation between OCT and CT measurements of the airway lumen and wall area were found (Coxson et al. 2008). OCT showed greater sensitivity in detecting changes in small airway wall measurements that relate to FEV_1 (Coxson et al. 2008). In addition, alveolar wall destruction in COPD can also be visualized with OCT. (See Figure 12.3.) There is also ongoing investigation and development of Doppler OCT for the quantitative assessment of blood flow and the microvasculature of the bronchial wall (Ohtani et al. 2012). This has significant potential for the evaluation of remodeling of airways vasculature in respiratory disease.

12.2.5 CONFOCAL FLUORESCENCE MICROENDOSCOPY AND ENDOCYSTOSCOPY

Techniques of *in vivo* microscopic evaluation of the bronchial mucosa have evolved to visualize cellular and extracellular structures in the airways. The first of these is confocal microendoscopy that utilizes a fibreoptic miniprobe via the working channel of a bronchoscope to deliver laser illumination to the tissue. One device is currently commercialized (Cell-viZio® Lung, Mauna Kea Technologies, Paris, France) and two different wavelengths are available, 488 nm and 660 nm (Thiberville et al. 2007, 2010). The miniprobe can be applied to the bronchial wall surface or advanced into a small peripheral bronchiole to image lung acini (Thiberville et al. 2009a, 2009b). The probe produces real-time images from a layer of 0–50 μm in depth below the surface with a lateral resolution of 3 μm and a circular field of view of 600 μm in diameter in 9–12 frames/second. The autofluorescence mode at 488 nm excitation produces detailed images of the bronchial subepithelial connective tissue and the alveolar structures (Thiberville et al. 2007, 2009b). In a series of 29 subjects, Thiberville and colleagues found that preinvasive bronchial lesions were associated with disorganization of the fibered network and reduction in fluorescence signal intensity but the technique could not distinguish between grades of disease (Thiberville et al. 2007; Thiberville et al. 2009a). In a subsequent series of 41 subjects and evaluation of alveoli, they were also able to detect increased alveolar macrophages in smokers and there was a strong correlation with cigarette consumption (Thiberville et al. 2009b). For more central airways, the excitation wavelength of 660 nm is used to visualize the bronchial epithelial cell layer after application of an exogenous fluorophore like methylene blue. This is required due to the low level of autofluorescence in the epithelial cell layer at 488 nm imaging (Thiberville et al. 2010, 2009a).

Confocal microendoscopy may be useful to expand our understanding of basement membrane remodeling in preinvasive and nonmalignant airway disease. It is presently limited by its inability to evaluate epithelial cells without an exogenous fluorophore.

In addition, tissue contact is required and resulting damage to the fragile epithelium can occur during bronchoscopy and associated cardiac/respiratory movement. It also has the potential to be useful in the evaluation of parenchymal lung disease and/or lesions if used in conjunction with navigational bronchoscopy techniques.

High magnification endocystoscopy (ECS) is another *in vivo* microscopic imaging technique that is undergoing investigation. Shibuya and colleagues described a prototype flexible endocystoscopy system (ECS XEC-300F, Olympus Optical Corp., Tokyo, Japan) that was 3.2 mm diameter and was inserted through the working channel of a large bronchoscope (Shibuya et al. 2011). The tip of the endocystoscope contained an optical magnifying lens system and CCD for imaging. It had a magnification of 570X and provided an observation depth of 0–30 μm, spatial resolution of 4.2 μm, and a field of 300 μm diameter (Shibuya et al. 2011). Following the application of topical 0.5% methylene blue, ECS imaging was performed on abnormal and normal areas identified by AFB and NBI during bronchoscopy in 22 high risk patients. ECS was able to produce images of cells similar to conventional histology (Shibuya et al. 2011). This is a promising technique that could obtain diagnostic microscopic evaluation of abnormal areas identified on bronchoscopy without biopsy. (See Figure 12.1.) At present, the size of the large bronchoscope limits access to some areas of the lung, particularly the upper lobes and the size of the endocystoscope limits access to smaller airways/parenchyma.

Therefore, the higher resolution of these two techniques makes it possible to image bronchial epithelial cells, subepithelial structures, and even alveoli in real-time. The use of exogenous agents to image epithelial cells remains a disadvantage. Other limitations include the requirement for direct tissue contact and the imaging is "end-on" without the ability to change the depth of the imaging plane.

12.2.6 LASER RAMAN SPECTROSCOPY

Raman spectroscopy (RS) is a technique that utilizes the Raman effect. This is an inelastic light scattering process that occurs when tissue is exposed to low-power laser light and a small proportion of incident photons are scattered with a corresponding change in frequency (Mahadevan-Jansen et al. 1996; Kaminaka et al. 2001; Huang et al. 2003; Short et al. 2008, 2011). The scattered light is collected for spectroscopic analysis. Raman spectra are shown by plotting the intensity of the scattered photons as a function of the frequency shift, that is, capturing a "fingerprint" of a specific molecular species (Mahadevan-Jansen et al. 1996; Huang et al. 2003). The difference between the incident and scattered frequencies corresponds to the vibrational modes of participating molecules (Mahadevan-Jansen et al. 1996). Light scattered from samples with different molecular compositions can be easily differentiated. This is a potentially powerful technique that can supply information at the molecular level.

The development of RS for clinical use has been limited until recently by two factors: the strong autofluorescence background of human tissue and the ability to measure spectra in a short time period to limit motion artifact. For clinical use in the lung, design of probes that can be used via the working channel of a bronchoscope and avoiding the issue of interference from Raman signals from the silica fiberoptics also needed exploration (Huang et al. 2003). The background tissue autofluorescence was found to be significantly reduced with the shift of the excitation wavelength into the near-infrared range (Kaminaka et al. 2001; Huang et al. 2003). The rapid acquisition of high-quality near-infrared Raman spectra (700–1800 cm^{-1}) in *ex vivo* lung specimens was shown to be possible (Huang et al. 2003). Subsequent work in a pilot study of 26 subjects by the same group showed that the use of near-infrared Raman spectroscopy *in vivo* during bronchoscopy using a 1.8-mm fiberoptic catheter was possible (Short et al. 2011). They were able to differentiate preinvasive lesions from benign tissue and reduce the false-positive rate of autofluorescence bronchoscopy (Short et al. 2011).

12.3 CONCLUSION

The technological development of optical imaging tools for incorporation into flexible bronchoscopy in recent years is exciting and encouraging. The capability to utilize and combine multiple imaging modalities, in a noninvasive or minimally invasive fashion, already exists and continues to expand to the microscopic level for the airway and the lung parenchyma. The adoption of these types of optical imaging techniques into bronchoscopic practice is likely to be the future path for the study and treatment of lung disease.

REFERENCES

Adams, K., Shah, P. L., Edmonds, L., and Lim, E. Test performance of endobronchial ultrasound and transbronchial needle aspiration biopsy for mediastinal staging in patients with lung cancer: Systematic review and meta-analysis. *Thorax* 2009; 64: 757–762.

Chhajed, P.N., Shibuya, K., Hoshino, H., Chiyo, M., Yasufuku, K., Hiroshima, K., and Fujisawa, T., A comparison of video and autofluorescence bronchoscopy in patients at high risk of lung cancer. *Eur Respir J* 2005; 25(6): 951–955

Chiyo, M., Shibuya, K., Hoshino, H., Yasufuku, K., Sekine, Y., Iizasa, T., Hiroshima, K., and Fujisawa, T. Effective detection of bronchial preinvasive lesions by a new autofluorescence imaging broncho videoscope system. *Lung Cancer* 2005; 48: 307–313.

Coxson, H., Quiney, B., Sin, D., Xing, L., McWilliams, A., Mayo, J., and Lam, S. Airway wall thickness assessed using computed tomography and optical coherence tomography. *Am J Respir Crit Care Med* 2008; 177: 1201–1206.

Coxson, H. O., Mayo, J., Lam, S. et al. New and current clinical imaging techniques to study chronic obstructive pulmonary disease. *Am J Respir Crit Care Med* 2009b; 180: 588–597.

Coxson, H. and Lam, S. Quantitative assessment of the airway wall using computed tomography and optical coherence tomography. *Proc Am Thorac Soc* 2009; 6: 439–443.

Edell, E., Lam, S., Pass, H. et al. Detection and localization of intraepithelial neoplasia and invasive carcinoma using fluorescence-reflectance bronchoscopy: An international, multicenter clinical trial. *J Thorac Oncol* 2009; 4: 49–54.

Fujimura, S., Sagawa, M., Saito, Y., Takahashi, H., Tanita, T., Ono, S., Matsumura, S., Kondo, T., and Sato, M. A therapeutic approach to roentgenographically occult squamous cell carcinoma of the lung. *Cancer* 2000; 89(11 Suppl.): 2445–2448.

Guardiola, E., Chaigneau, L., Villanueva, C., and Pivot, X. Is there still a role for triple endoscopy as part of staging for head and neck cancer? *Curr Opin Otolaryngol Head Neck Surg* 2006; 14: 85–88.

Häussinger, K., Stanzel, F., Huber, R.M., Pichler, J., and Stepp, H. Autofluorescence detection of bronchial tumors with the D-Light/AF. *Diagn Ther Endosc* 1999; 5: 105–112.

Herth, F., Eberhardt, R., Anantham, D., Gompelmann, D., Zakaria, M., and Ernst, A. Narrowband imaging bronchoscopy increases the specificity of bronchoscopic early lung cancer detection. *J Thorac Oncol* 2009; 4: 1060–1065.

Huang, Z., McWilliams, A., Lui, H., McLean, D., Lam, S., and Zeng, H. Near-infrared Raman spectroscopy for optical diagnosis of lung cancer. *Int J Cancer* 2003; 107: 1047–1052.

Ikeda, S., Tsuboi, E., Ono, R., and Ishikawa, S. Flexible broncho fiberscope. *Jpn J Clin Oncol* 1971; 1: 55–65.

Ikeda, N., Honda, H., Hayashi, A., Usuda, J., Kato, Y., Tsuboi, M., Ohira, T., Hirano, T., Kato, H., Serizawa, H., and Aoki, Y. Early detection of bronchial lesions using newly developed videoendoscopy-based autofluorescence bronchoscopy. *Lung Cancer* 2006; 52: 21–27.

Kaminaka, S., Yamazaki, H., Ito, T., Kohda, E., and Hamaguchi, H. Near-infrared Raman spectroscopy of human lung tissues: Possibility of molecular-level cancer diagnosis. *J Raman Spectrosc* 2001; 32: 139–141.

Keith, R., Miller, Y., Gemmill, R., Drabkin, H., Dempsey, E., Kennedy, T., Prindiville, S., and Franklin, W. Angiogenic squamous dysplasia in bronchi of individuals at high risk for lung cancer. *Clin Cancer Res* 2000; 6: 1616–1625.

Kennedy, T.C., McWilliams, A., Edell, E., Sutedja, T., Downie, G., Yung, R., Gazdar, A., Mathur, P.N. American College of Chest Physicians. Bronchial intraepithelial neoplasia/early central airways lung cancer: ACCP evidence-based clinical practice guidelines (2nd edition). *Chest* 2007 Sep; 132(3 Suppl.): 221S–233S.

Kurimoto, N., Murayama, M., Yoshioka, S., Nishisaka, T., Inai, K., Dohi, K. Assessment of usefulness of endobronchial ultrasonography in determination of depth of tracheobronchial tumor invasion. *Chest* 1999; 115(6): 1500–1506.

Kurimoto, N., Murayama, M., Yoshioka, S., and Nishisaka, T. Analysis of the internal structure of peripheral pulmonary lesions using endobronchial ultrasonography. *Chest* 2002; 122: 1887–1894.

Lam, S., MacAulay, C., Hung, J. et al. Detection of dysplasia and carcinoma *in situ* with a lung imaging fluorescence endoscope device. *J Thorac Cardiovasc Surg* 1993; 105: 1035–1040.

Lam, S. The role of autofluorescence bronchoscopy in diagnosis of early lung cancer. In:Hirsch, F.R., Bunn, P.A., Jr., Kato, H. et al. (eds.), *IASLC Textbook for Prevention and Detection of Early Lung Cancer*. London: Taylor & Francis, 2006; 149–158.

Lam, S., Standish, B., Baldwin, C. et al. *in vivo* optical coherence tomography imaging of preinvasive bronchial lesions. *Clin Cancer Res* 2008; 14: 2006–2011.

Lee, P., Brokx, H.A.P., Postmus, P.E., and Sutedja T. Dual digital video-autofluorescence imaging for detection of preneoplastic lesions. *Lung Cancer* 2007; 58: 44–49.

Lee, P., van den Berg, R.M., Lam, S., Gazdar, A., Grunberg, K., McWilliams, A., LeRiche, J., Postmus, P., and Sutedja, T. Color fluorescence ratio for detection of bronchial dysplasia and carcinoma *in situ*. *Clin Cancer Res* 2009; 15: 4700–4705.

McWilliams, A.M., Mayo, J.R., Ahn, M.I., MacDonald, S.L.S., and Lam, S. Lung cancer screening using multi-slice thin-section computed tomography and autofluorescence bronchoscopy. *J Thorac Oncol* 2006; 1(1): 61–68.

Mahadevan-Jansen, A. and Richards-Kortum, R. Raman spectroscopy for the detection of cancers and precancers. *J Biomed Opt* 1996; 1: 31–70.

Nakamura, H., Kawasaki, N., Hagiwara, M., Ogata, A., Saito, M., Konaka, C., and Kato, H. Early hilar lung cancer-risk for multiple lung cancers and clinical outcome. *Lung Cancer* 2001; 33(1): 51–57.

The National Lung Screening Trial Research Team. Reduced lung cancer mortality with low dose computed tomographic screening. *N Engl J Med* 2011; 365: 395–409.

Ohtani, K., Lee, A., and Lam, S. Frontiers in bronchoscopic imaging. *Respirology*, 2012; 17: 261–269.

Shibuya, K., Hoshino, H., Chiyo, M., Yasufuku, K., Iizasa, T., Saitoh, Y., Baba, M., Hiroshima, K., Ohwada, H., and Fujisawa, T. Subeptihelial vascular patterns in bronchial dysplasias using a high magnification bronchovideoscope. *Thorax* 2002; 57: 902–907.

Shibuya, K., Hoshino, H., Chiyo, M., Iyoda, A., Yoshida, S., Sekine, Y., Iizasa, T., Saitoh, Y., Baba, M., Hiroshima, K., Ohwada, H., and Fujisawa, T. High magnification bronchovideoscopy combined with narrow band imaging could detect capillary loops of angiogenic squamous dysplasia in heavy smokers at high risk for lung cancer. *Thorax* 2003; 58: 989–995.

Shibuya, K., Nakajima, T., Fujiwara, T., Chiyo, M., Hoshino, H., Moriya, Y., Suzuki, M., Hiroshima, K., Nakatani, Y., and Yoshino, I. Narrow band imaging with high-resolution bronchovideoscopy: A new approach for visualizing angiogenesis in squamous cell carcinoma of the lung. *Lung Cancer* 2010; 69: 194–202.

Shibuya, K., Fujiwara, T., Yasufuku, K., Mohamed Alaa, R.M., Chiyo, M., Nakajima, T., Hoshino, H., Hiroshima, K., Nakatani, Y., and Yoshino, I. *In vivo* microscopic imaging of the bronchial mucosa using an endo-cystoscopy system. *Lung Cancer* 2011; 72: 184–190.

Short, M., Lam, S., McWilliams, A., Zhao, J., Lui, H., and Zeng, H. Development and preliminary results of an endoscopic Raman probe for potential *in vivo* diagnosis of lung cancers. *Opt Lett* 2008; 33: 711–713.

Short, M., Lam, S., McWilliams, A., Ionescu, D., and Zeng, H. Using laser Raman spectroscopy to reduce false positives of autofluorescence bronchoscopy: A pilot study. *J Thorac Oncol* 2011; 6: 1206–1214.

Steinfort, D.P., Khor, Y.H., Manser, R.L. et al. Radial probe endobronchial ultrasound for the diagnosis of peripheral lung cancer: Systematic review and meta-analysis. *Eur Respir J* 2011; 37: 902–910.

Sun, J., Garfield, D.H., Lam, B. et al. The value of autofluorescence bronchoscopy combined with white light bronchoscopy compared with white light alone in the diagnosis of intraepithelial neoplasia and invasive lung cancer: A meta-analysis. *J Thorac Oncol* 2011; 6(8): 1336–1344.

Swensen, S.J., Jett, J.R., Hartman, T.E. et al. CT screening for lung cancer: Five-year prospective experience. *Radiology* 2005; 235: 259–265.

Takahashi, H., Sagawa, M., Sato, M., Sakurada, A., Endo, C., Ishida, I., Oyaizu, T., Nakamura, Y., and Kondo, T. A prospective evaluation of transbronchial ultrasonography for assessment of depth of invasion in early bronchogenic squamous cell carcinoma. *Lung Cancer* 2003; 42: 43–49.

Thiberville, L., Moreno-Swirc, S., Vercauteren, T., Peltier, E., Cave, C., and Bourg-Heckly, G. *In vivo* imaging of the bronchial wall microstructure using fibered confocal fluorescence microscopy. *Am J Respir Crit Care Med* 2007; 175: 22–31.

Thiberville, L., Salaun, M., Lachkar, S., Dominique, S., Moreno-Swirc, S., Vever-Bizet, C., Bourg-Heckly, G. Confocal fluorescence endomicroscopy of the human airways. *Proc Am Thorac Soc* 2009a; 6: 444–449.

Thiberville, L., Salaun, M., Lachkar, S., Dominique, S., Moreno-Swirc, S., Vever-Bizet, C., Bourg-Heckly, G. Human *in vivo* fluorescence microimaging of the alveolar ducts and sacs during bronchoscopy. *Eur Respir J* 2009b; 33: 974–985.

Thiberville, L., Salaun, M.. Bronchoscopic advances: On the way to the cell. *Respiration* 2010; 79: 441–449.

Tsuboi, M., Hayashi, A., Ikeda N. et al. Optical coherence tomography in the diagnosis of bronchial lesions. *Lung Cancer* 2005; 49: 387–394.

Usuda, J., Ichinose, S., Ishizumi, T., Hayashi, H., Ohtani, K., Maehara, S., Ono, S., Kajiwara, N., Uchida, O., Tsutsui, H., Ohira, T., Kato, H., and Ikeda, N. Management of multiple primary lung cancer in patients with centrally located early cancer lesions. *J Thorac Oncol* 2010; 5(1): 62–68.

Vincent, B., Fraig, M., and Silvestri, G. A pilot study of narrow-band imaging compared to white light bronchoscopy for evaluation of normal airways and premalignant and malignant airways disease. *Chest* 2007; 131: 1794–1799.

Wagnieres, G., McWilliams, A., and Lam, S. Lung cancer imaging with fluorescence endoscopy. In: Mycek, M., Pogue, B. (eds.), *Handbook of Biomedical Fluorescence*. New York: Marcel Dekker, 2003; 361–396.

Woolner, L., Fontana, R.S., Cortese, D.A. et al. Roentgeno graphically occult lung cancer: Pathologic findings and frequency of multicentricity during a 10-year period. *Mayo Clinic Proc* 1984; 59: 453–466.

Zeng, H., Petek, M., Zorman, M.T., McWilliams, A., Palcic, B., and Lam, S. 2004. Integrated endoscopy system for simultaneous imaging and spectroscopy for early lung cancer detection. *Opt Lett* 29: 587–589.

13 Clinical Applications in the Gastrointestinal Tract

Beau Standish
Ryerson University, Toronto, Ontario, Canada

Victor Yang
Ryerson University and Sunnybrook Health
Sciences Centre, Toronto, Ontario, Canada

Naoki Muguruma and Norman Marcon
St. Michael's Hospital, Toronto, Ontario, Canada

CONTENTS

13.1 INTRODUCTION

The major thrust of this chapter is the adaptation of endoscopic technologies to improve the detection of precancerous lesions at a stage when they are still limited to the mucosa and curable by endoscopic means or minimally invasive surgery. The gastrointestinal (GI) cancer burden worldwide is immense. In 2002 there were 3.2 million new cases of GI cancer resulting in 2.4 million deaths (Ferlay et al. 2004). It is physically and economically impossible to endoscope everyone, who, for a variety of geographic and genetic causes, would be at risk to develop GI malignancy. What is desperately needed is a biomarker or a panel of biomarkers to select appropriate patients at risk who would benefit from endoscopic interrogation. Although this utopian goal is desirable, it is not yet attainable.

However, progress and innovation have occurred in endoscopic technology development, where different platforms are vying for "prime time" and the potential opportunity to revolutionize clinical practice. These technological innovations are a result of productive collaborations among gastroenterologists, physicists, engineers, industrial partners, and are exemplified in many of the multidisciplinary research groups active in gastroenterology. While white light endoscopy remains the standard for endoscopic examination, technological advances in imaging or spectroscopic techniques are at different stages of the "bench-top" to "bedside" pathway leading from preclinical study to patient care.

This chapter reviews selected state-of-the-art endoscopic modalities, their present, and future clinical applications, and summarizes the existing evidence based literature on how these innovative, optically-based methods may improve the precancerous detection rates for patients undergoing endoscopic screening.

13.2   WHITE LIGHT ENDOSCOPY

Since the inception of the electronic endoscope by Welch-Allyn Inc., clinicians have been able to observe the gastrointestinal tract with high quality images in real-time. The endoscope has become the tool most commonly associated with gastroenterologists. This expanding technology in endoscopes has not only resulted in improved diagnostic capability but has been the platform for the ever-expanding field of therapeutic endoscopy. In spite of evolving improvements in technology, the fundamental issue resides in the ability of the endoscopist to perform a careful examination based on the close observation of the topographic geography of the mucosa (i.e., lumps, bumps, color changes), knowledge of risk conditions, and experience. This diagnostic algorithm must also include histological examination of random or targeted biopsies.

The switch from fiberoptic-based endoscopic imaging to chip-based endoscopic imaging, has resulted in significant improvement in the clarity of mucosal detail. This is related to the manufacture of CCD (Charged Coupled Device) chip technology that is continually evolving and can currently produce an image in the range of 850,000 to greater than 1 million pixels (Kwon et al. 2009). These high pixel density devices fulfill the definition of high resolution and high definition (HD). Intimately involved with the chip technology is the availability of HD display monitors that have enabled the improvement of the image quality displayed during the endoscopic examination.

As a result of improved image clarity due to high-resolution chip technology, the addition of other multimodal optical technologies, such as narrow-band imaging, autofluorescence, etc., will have a greater challenge to prove their worth. The question remains if the addition of these new image-enhancing technologies will improve upon the careful examination of an observant endoscopist, with a good understanding of predisposing conditions and low-tech "old fashioned" dye spraying. The clinical benefit of these evolving multimodal technologies will be ultimately determined by the design of appropriate evidence-based trials that will demonstrate cost effectiveness in predicting histology and clinical outcomes.

13.3 RAMAN SPECTROSCOPY

Raman spectroscopy is an optical technique that can provide detailed information about the chemical composition and molecular structure of biological tissues, where each molecule emits a unique spectral signature or "fingerprint" specific to its tissue microenvironment. It has been demonstrated that as normal tissues undergo transformation towards cancer, their Raman spectral signatures change to reflect specific stages of carcinogenesis. This capability is especially appealing for detecting molecular-level changes in tissue for the purpose of endoscopic detection of early cancer. The exploitation of these molecular differences between normal and dysplastic mucosa in the gastrointestinal tract may allow Raman spectroscopy to assist the endoscopist in identifying white light-occult dysplastic lesions in "wide-field disease" such as Barrett's esophagus and chronic ulcerative colitis. However, as a point-detection technique, Raman spectroscopy is somewhat impractical as the area of interest must still be identified by the endoscopist through additional means (i.e., topography, change in color, NBI, etc.), at which time areas of high suspicion can be interrogated with Raman spectroscopy and confirmed by biopsy.

Typically, Raman spectroscopy is performed by illuminating tissue, using a fiber optic probe, with monochromatic near-infrared (NIR) light that is absorbed and inelastically scattered by the unique vibrational/rotational modes of molecular bonds associated with a variety of tissue components (i.e., mucosal proteins, lipids, and nucleic acids). The earliest work in this area related to gastroenterology began by studying the complex Raman spectra of *ex vivo* tissues obtained during routine endoscopy (Shim and Wilson 1997). However, over the past decade, several investigators have applied Raman spectroscopy in different technical configurations to acquire *in vivo* Raman spectra (Shim et al. 1999; Bakker Schut et al. 2000; Kendall et al. 2003; Wang and Van Dam 2004) for the purpose of detecting dysplasia *in vivo*. For example, Shim et al. designed and built a near-infrared (NIR)-based fiber optic probe compatible with a standard colonoscope for *in vivo* Raman spectroscopy measurements (Shim and Wilson 1997). Briefly, this was achieved with an optically filtered, flexible, fiber optic probe (~2 mm diameter) capable of "beam-steering" that was passed down the endoscope's instrument channel and placed in gentle contact with the tissue surface (Shim et al. 1999). The quality of measured signal was optimized by minimizing distortion related to excessive pressure on the probe tip and the angle of placement in relation to tissue surface. It was observed that the spectra from normal and diseased tissues revealed only subtle differences. This early data demonstrated the need for amplification of signals and better algorithms for differentiation and classification between normal and diseased tissue.

Building upon previous work, Molckovsky et al. (2003) studied colon polyps from eight patients, where 54 *ex vivo* Raman spectra were analyzed (20 hyperplastic, 34 adenomatous). The spectral-based diagnostic algorithms identified adenomatous polyps with 91% sensitivity, 95% specificity, and 93% accuracy. *In vivo* adenomas (10) were distinguished from hyperplastic polyps (n = 9) with 100% sensitivity, 89% specificity, and 95% accuracy. They concluded that NIR Raman spectroscopy differentiated adenomatous from hyperplastic polyps with high diagnostic accuracy.

New research involving surface enhanced Raman scattering (SERS) may potentially improve the detection sensitivity by up to 6–10 orders of magnitude over conventional Raman spectroscopy (Astorga-Wells et al. 2005; Yang, Jenkins et al. 2005; Dittrich, Tachikawa, and Manz 2006). Disappointingly, there have been limited reports on the clinical application of this technology as it is in its earliest stages of development with little advancement over the last decade.

13.4 LIGHT-SCATTERING SPECTROSCOPY

Light-scattering spectroscopy (LSS) provides morphological information about subcellular features such as nuclei and mitochondria from elastically scattered visible light (Backman et al. 2000). The pathological hallmarks of dysplasia are increased nuclear size and crowding. The potential use of using LSS *in vivo* to detect epithelial nuclear crowding and enlargement in Barrett's esophagus was evaluated by Wallace et al. (2000). Diffusely reflected white light was spectrally analyzed to obtain the size distribution of cell nuclei in the colonic mucosal layer, from which the percentage of enlarged nuclei and the degree of crowding were determined. Dysplasia was assigned if more than 30% of the nuclei exceeded 10 μm in diameter. This automated classification was compared to the histological findings of four pathologists who were blinded to the light-scattering assessment. Using this threshold, the sensitivity and specificity for detecting low-grade dysplasia (LGD) and high-grade dysplasia (HGD) were both 90%, which demonstrates the potential of this method to detect LGD and HGD in Barrett's esophagus.

A more recent study by Lovat et al. assessed the potential for elastic LSS to detect HGD or cancer within Barrett's esophagus in 81 patients (Lovat et al. 2006). Measurements collected *in vivo* were matched with histological specimens taken from the same sites within the Barrett's esophagus field, and biopsies were classified as either "low risk" (nondysplastic or LGD) or "high risk" (HGD or cancer). Elastic scattering spectroscopy detected high-risk sites with 92% sensitivity and 60% specificity, and differentiated high-risk sites from inflammation with 79% sensitivity and 79% specificity. This group concluded that if LSS was used to target biopsies during endoscopy, the number of low-risk biopsies taken would decrease by ~60% with minimal loss of accuracy. Furthermore, a negative spectroscopic result would exclude HGD or cancer with an accuracy of >99.5%. The authors suggested that LSS has the potential to target conventional biopsies in Barrett's surveillance, which would save significant time for the endoscopist and the pathologist while, presumably, increasing cost effectiveness.

To date, the potential of using LSS *in vivo* in the human colon has been reported in one clinical study by Dhar et al. (2006). They used the elastic light scattering spectra obtained from 138 sites in 45 patients undergoing colonoscopy. These spectra were then compared with conventional biopsy specimens taken from the same site, including normal colonic mucosa, hyperplastic polyps, adenomatous polyps, chronic colitis, and colon cancer. Spectral analysis was carried out with a validated computerized model that used principal component analysis followed by linear discriminant analysis. The sensitivity and the specificity of differentiating adenomatous from hyperplastic polyps were 84 and 84%; for cancer from adenomatous polyps, 80 and 75%; for colitis from normal tissue, 77 and 82%, respectively; and dysplastic mucosa (in polyps) from colitis, 85 and 88%, respectively.

LSS has been explored as one possible technology for clinical diagnosis of dysplasia. However, the point detection aspect is a major drawback, and will likely prevent it from being accepted in mainstream clinical practice, despite the potential to detect subcellular tissue changes associated with early cancer. Future work in this area will need to include a concerted effort between basic scientists; commercial companies and validation by multicenter trials to further develop this technology and its associated diagnostic algorithms for clinical use.

13.5 IMAGE ENHANCING ENDOSCOPY

13.5.1 CHROMOENDOSCOPY

Chromoendoscopy is a widespread and simple technique for enhancement of subtle mucosal abnormalities (Wong Kee Song et al. 2007). Agents used for chromoendoscopy are classified into three subgroups in regards to characteristics of the dye: (1) metabolic or vital (i.e., Lugol's or Methylene blue), (2) contrast agents (i.e., Indigo Carmine), and (3) chemical mucosal enhancement (i.e., acetic acid). The technique involves spraying a solution (dye) on the mucosa, via the accessory channel or using a special catheter, with fine holes at the distal tip to enable an even spray. Prior to performing chromoendoscopy, especially in the stomach, the mucosal surface should be sprayed with a 10% acetylcysteine solution to wash away superficial mucus. This pretreatment is necessary to prime the absorbing cells for successful uptake of the dye, which enhances tissue contrast, defines the lesion border, and improves topographic characterization during the procedure. In conjunction with high-resolution and magnification endoscopy, chromoendoscopy results in better contrast especially where there are either subtle elevations or depressions, resulting in improved macroscopic analysis (Jung and Kiesslich 1999).

Dye spraying with Lugol's solution (0.5–3.0%) is the classic success story of an inexpensive, readily available dye that has been shown to be effective in identifying early squamous cell cancer in the esophagus. Lugol's is a mixture of potassium iodide and iodine. Lugol's reacts with glycogen in nonkeratinized squamous epithelium and results in a dark green-brown color as the normal esophageal mucosa is richly endowed with glycogen. Dysplastic areas, deficient in glycogen, are not stained and are therefore readily observed as pale yellow islands described as Lugol's negative areas (see Figure 13.1). False positives occur in areas of dense inflammation. Care must be exercised with Lugol's solution because it can cause mucosal irritation, retrosternal pain, and is particularly dangerous if aspirated into the trachea. Excess solution must be frequently aspirated from the esophagus and fundus. A solution of sodium thiosulfate can be sprayed into the esophagus to neutralize the irritant effect

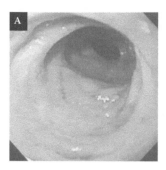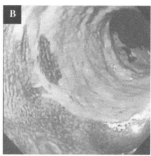

FIGURE 13.1 (See color insert) Lugol's solution used to detect squamous cell carcinoma. A: White light endoscopic view of esophagus. B: Endoscopic view following staining with Lugol's solution showing squamous cell carcinoma in the area that is unstained (yellow). The area stained dark greenish-brown is normal squamous mucosa.

of the free iodine content of Lugol's. This solution, while commonly available in Japan, is not available in North America (Kondo et al. 2001).

Lee et al. have recently reported on the early detection of mucosal high-grade squamous dysplasia and invasive cancer via Lugol's solution and NBI (Lee et al. 2009). A total of 54 endoscopic sessions (44 patients) evaluated the screening performance of Lugol's chromoendoscopy and resulted in a sensitivity, specificity, positive predict value (PPV), negative predict value, and accuracy of 88.9%, 72.2%, 61.5%, 92.9%, and 77.8%, respectively. Additionally, Lee et al. screened the same patients with narrow band imaging (NBI, discussed in detail in Section 13.4.2), obtaining the following results: 88.9%, 97.2%, 94.1%, 94.6%, and 94.4%, respectively. The authors concluded that the Lugol's staining is unsurpassed for delineating the margins of cancerous tissue to facilitate targeted biopsy or mucosectomy, while NBI provides an additional imaging contrast to overcome the low specificity and low PPV associated with Lugol's staining to improve targeted biopsy. Therefore, the combination of both detection modalities provides complimentary information to maintain high accuracy, while mitigating the overall risk of obtaining false-negative results.

Methylene blue (0.5% dye solution) is a vital stain that is actively absorbed by the epithelial cells of the small bowel and the colon. Intestinal metaplasia, having characteristics resembling small bowel epithelium, also absorbs methylene blue, whereas normal gastric mucosa and squamous epithelia remain unstained. These features may be helpful in detecting intestinal metaplasia in columnar-lined epithelium suspicious for Barrett's esophagus. However, it does not seem to be useful in detecting dysplasia in Barrett's. Although categorized as an absorptive stain, methylene blue is also used by some endoscopists in North America and Europe more commonly as a contrast stain, in particular because of the lower costs as compared with indigo carmine (Peitz and Malfertheiner 2002). There have been no significant side effects associated with methylene blue use, although there has been unsubstantiated speculation on the potential of methylene blue inducing oxidative cellular damage when photosensitized by white light.

Indigo carmine, a contrast dye (0.1%–0.4% solution with sterile water) that pools in mucosal crevices without any cellular staining, has been used for diagnosis of flat early gastric and colorectal lesions to enhance the visualization of the lesion.

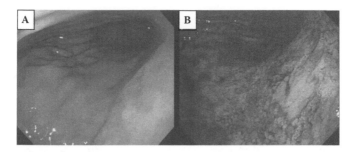

FIGURE 13.2 (See color insert)   A: Conventional white light endoscopy of the stomach showing an irregular mucosal pattern with mild erythema. B: Chromoendoscopy with acetic acid indigocarmine mixture (AIM) showing the border of the lesion clearly and biopsy from this area showed tubular adenocarcinoma. (Courtesy of Koichi Okamoto, Tokushima University Hospital, Japan.)

Most of its reported use has been to detect diminutive, flat and depressed lesions in the colon using magnification endoscopes. Recently, an acetic acid-indigocarmine mixture (AIM) was reported to be useful for identifying the margin of early gastric neoplasia when compared to indigo carmine alone. AIM combines 0.6% acetic acid and 0.4% indigocarmine into one solution and is sprinkled, at low pressure, onto the gastric mucosa by means of a syringe attached to the accessory channel of the endoscope. Kawahara et al. (2009) have demonstrated the diagnostic accuracy of the AIM solution in detecting the margins of early gastric cancers to be 90.7%. This is an impressive improvement over white light (50.0%) and indigo carmine (75.9%) accuracy, as AIM is also easy to use, safe, and inexpensive. An example of an enhanced visualization of gastric cancer, using the AIM solution, can be viewed in Figure 13.2.

Acetic acid has been routinely used since the inception of colposcopy as its topical application in the cervix results in an increase in the opacity of the mucosal surface. This opacity is associated with atypical dysplastic epithelium, which guides the gynecologist to target biopsy when screening for cervical intraepithelial neoplasia. Guelrud and Herrera originally adapted this technique in 1998 to aid in the detection of residual Barrett's in patients who had undergone ablation therapy (Guelrud and Herrera 1998). Since that time there have been no large controlled validation studies, however, several reports have described that careful examination of the mucosal pattern, after the application of acetic acid (~1.5% solution) in combination with a magnifying endoscopy, can improve the diagnostic accuracy in Barrett's esophagus (Hoffman et al. 2006), gastric neoplasia (Tanaka et al. 2006) and colon polyp (Togashi et al. 2006). The advantages of this technique are that it is low-cost, low-volume (reduces risk of aspiration) and its effect on tissue is transient, allowing for multiple applications. This chemical mucosal enhancing method is well known and widely used in the endoscopy community, yet there does not exist an agreed-upon protocol for the classification and/or staging of mucosal pathologies.

Although the biological mechanisms of chromoendoscopy may not be fully understood, its advantages are that the dyes are inexpensive, safe, and have established a long clinical track record. Whether the new optical filter or electronic chromoendoscopy will be as good as or better remains to be seen.

13.5.2 Narrow Band Image

Narrow band imaging (NBI), sometimes referred to as virtual chromoendoscopy, is an innovative optical-digital technology that visualizes the microvascular structure and epithelial crests (mucosal pattern; Gono et al. 2003) of the gastrointestinal tract. NBI uses optimized filters with bandwidths of ~30 nm and central wavelength at ~415 nm (blue) and ~540 nm (green), that enhance the microvascular and texture pattern on the mucosal surface, without the need for dye spraying. The technique is based on the fact that shorter wavelengths (i.e., blue) will only penetrate the superficial region of the mucosa (~200 microns), where the capillary network resides. This capillary network is used as a marker for dysplatic transformation as the size, density, and tortuosity of these vessels increases during the neovascularization process associated with the transformation from normal to diseased tissue. This visual enhancement by NBI can be viewed as an increase in the intensity of the color brown and red flagged by a trained endoscopist. Through an on/off switch located on the handle of the endoscope, one can activate the NBI image instantly (Figures 13.3–13.4).

Wada et al. recently reported on the feasibility of using NBI to diagnose colorectal lesions. It was assumed that a faint vascular pattern was diagnostic for hyperplastic polyps, allowing for differentiation between neoplastic and nonneoplastic lesions with a sensitivity of 90.9% and a specificity of 97.1%. When irregular and sparse vascular patterns were observed it was assumed to be a "red flag" for invasive submucosal cancer. This classification resulted in a sensitivity of 100%, specificity of 95.8%, and an accuracy rate of 96.1% (Wada et al. 2009).

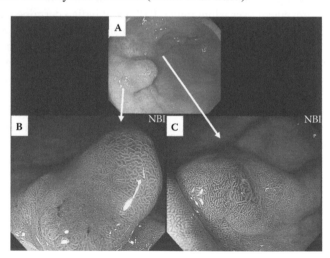

FIGURE 13.3 (See color insert) A: A couple of polypoid lesions were identified in the body of the stomach with WLE. B: NBI with magnification image showing smaller mucosal pattern in the central depressed area comparing to the surrounding area and the depressed area has abundant microvessels. Biopsy from the central area showed differentiated tubular adenocarcinoma. C: NBI with magnification image showing regular oval mucosal pattern and biopsy showed no malignant cell, which proved this as a scar due to the previous endoscopic resection. (Courtesy of Atsuo Oshio, National Hospital Organization, Kochi Hospital, Japan.)

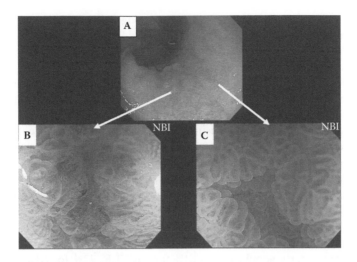

FIGURE 13.4 (See color insert) A: Reddish depressed lesions were identified in the body of the stomach with white light endoscopy. B: NBI with magnification image showing irregular microvascular pattern and biopsy showed differentiated tubular adenocarcinoma. C: NBI with magnification image showing regular microvascular pattern and biopsy showed focal chronic gastritis. (Courtesy of Atsuo Oshio, National Hospital Organization, Kochi Hospital, Japan.)

On the other hand, some reports suggest that NBI has failed to demonstrate an improved detection of adenoma in the colon. Rex et al (2007) conducted a randomized controlled trial of WLE and NBI for the purpose of finding adenomatous polyps in 434 patients. A single experienced endoscopist performed all endoscopies, where withdrawal of the endoscope was randomized to either WLE or NBI. There was no significant difference in adenoma detection in both arms, although the time of withdrawal was slightly longer in the NBI arm.

Three recent publications (Adler et al. 2008; Adler, Aschenbeck et al. 2009; Paggi et al. 2009) also concur that the addition of NBI did not demonstrate significant differences in the adenoma detection rate. However, they found NBI with magnification does improve the ability to differentiate diminutive polyps as to whether they are hyperplastic or adenomatous.

Two studies (Ponchon et al. 2007; Lee et al. 2009) suggest that NBI may be as sensitive as Lugol's solution for the detection of superficial squamous cell cancer. Given the practical disadvantages of Lugol staining (use of spray catheters, risk of aspiration for hypopharyngeal and proximal esophageal imaging, iodine allergy, and esophageal spasm), NBI seems to be a valid alternative to Lugol's and indeed may also be complimentary.

The NBI technology has also been studied for its potential use in Barrett's esophagus with contradictive findings. In a prospective cohort study, Sharma et al. (Sharma et al. 2006) assessed the potential of NBI in predicting histology in 51 patients undergoing surveillance for BE. NBI images were graded according to mucosal (ridge/villous, circular and irregular/distorted) and vascular (normal/abnormal) pattern. All samples were correlated with histology in a blinded manner. The sensitivity, specificity, and positive predictive value of the ridge villous pattern for predicting intestinal metaplasia alone were 93.5%, 86.7%, and 94%. As predictors of HGD, the sensitivity, specificity, and positive predictive value of irregular distorted pattern and irregular vascular pattern were 100%, 98.7%, and 93.3% and 100%, 97.2%, and 94.7%, respectively. It should be noted that these observations required not only high resolution endoscopes, but also magnification. These endoscopes are not commonly available in the vast majority of endoscopic units.

A number of additional studies using NBI in Barrett's esophagus, each proposing alternative classification systems, demonstrate promising results with both high sensitivities and specificities for the differentiation nondysplastic BE from HGD/cancer (Kara, Ennahachi et al. 2006; Goda et al. 2007). In contrast, a more current study by Curves et al. (Curvers, Bohmer et al. 2008) evaluated the intra-observer agreement and the additional benefit of NBI over high resolution (HR) WLE. Five experts in the field of Barrett's and seven nonexpert endoscopists independently evaluated still images from 50 areas obtained with HRWLE and NBI. The overall yield for correctly identifying early neoplasia was 81% for HRWLE, 72% for NBI, and 83% for HRWLE + NBI. The authors concluded that the addition of NBI did not improve the yield for identifying neoplasia in Barrett's esophagus.

Studies investigating the utility of NBI in the stomach are limited. Uedo et al. validated a proposed classification that distinguished the presence of gastric intestinal metaplasia displayed as a light blue epithelial crest on NBI magnification imaging (Uedo et al. 2006). First, histological correlation was obtained from 44 biopsies taken from the light blue crest area and 44 control biopsies from 34 patients with atrophic gastritis. The sensitivity and specificity for the light blue crest appearance was 91% and 84%, respectively. These findings were validated in a large prospective study of 107 consecutive patients. The epithelial appearance of a light blue crest correlated with histological presence of intestinal metaplasia with a sensitivity of 89%, a specificity of 93%, a PPV of 92%, and a NPV of 91%. The authors concluded that a light blue crest visualized, using magnification NBI on the gastric mucosa is highly accurate for the detection of gastric intestinal metaplasia.

Nakayoshi et al. prospectively correlated the vascular pattern of NBI magnification images and histologic findings in 165 patients with early gastric cancer (Nakayoshi et al. 2004). The microvascular patterns were classified into three groups: A-fine network, B-corkscrew, and C-unclassified. Among the 109 patients with differentiated adenocarcinoma, a fine network microvascular pattern was observed in 72 cases (66.1%) and among the 56 with undifferentiated gastric adenocarcinoma the corkscrew pattern was observed in 48 cases (85.7%, $P = 0.0011$). The authors concluded that magnification NBI is not sufficient for diagnosis, but can be used to predict some histological characteristics of gastric cancer and might allow for improved differentiation between benign and malignant minute lesions.

The attempt to establish a role for NBI in the early detection of gastrointestinal lesions has had controversial results. However, it has led to increased attention to the value of careful endoscopic observation. NBI has spawned a great deal of interest in the early detection of dysplastic lesions. It has resulted in a better understanding of mucosal features and should lead to useful clinical classification and treatment algorithms.

13.5.3 ALTERNATIVE ELECTRONIC IMAGE ENHANCEMENT AND POST-IMAGE PROCESSING

Two recent commercial endoscopic image-enhancing technologies have become available, the FUJI Intelligent Color Enhancement (FICE or optimal band imaging) and more recently "i-scan" by Pentax. These techniques produce digital images through electronic post-processing image enhancement methods to provide contrast of mucosal tissue structures. Publications are limited and comparative studies are nonexistent. Osawa et al. have recently published (Osawa et al. 2008) on optimal band imaging and were able to enhance the identification of the demarcation line of depressed-type early gastric cancer without the need of mechanical magnification in 26 of 27 cases (96%). It is likely that as this post-processing technology penetrates the marketplace; further studies will be published using FICE or i-scan to replicate NBI findings.

13.5.4 AUTOFLUORESCENCE IMAGING

In contradistinction of point spectroscopy, autofluorescence imaging (AFI) allows visualization of the entire endoscopic mucosal field at once. AFI takes advantage of specific endogenous molecules that when excited by a short-wavelength light (i.e., blue light), they emit a longer wavelength fluorescence light. These changes in wavelength correlate with histological alteration. The fluorescent signal can be selectively detected and examined through specialized optical filters and spectral analyzers. Due to the changes of endogenous fluorophores in dysplastic tissue, the altered autofluorescence spectrum is translated into a pseudo-color image. The color composition is dependent on the specific system used. In the Olympus video AFI system, nondysplastic tissue appears green and dysplastic tissue appears blue or violet. It has previously been shown that AFI can distinguish differences between high-grade intraepithelial neoplasia (HGIN) and nondysplastic Barrett's mucosa (Figure 13.5) based on variations in the optical intensity and the spectral composition of the reflected autofluorescence signal (Izuishi et al. 1999; Georgakoudi et al. 2001).

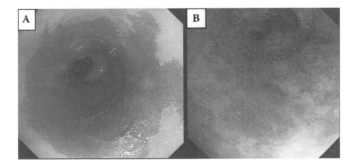

FIGURE 13.5 (See color insert)   A: High resolution WLE image of the long segment of Barrett's esophagus. B: AFI image showing a patch of positive fluorescence (violet) at 6 o'clock. Biopsy from this area showed high-grade dysplasia. (Courtesy of L.M. Wong Kee Song, Mayo Clinic, Rochester, Minnesota.)

Kara et al. published a study in which 47 suspicious lesions were detected with AFI, where only 28 lesions actually contained HGIN resulting in a false-positive rate of 40% (Kara, Peters et al. 2006). This led to the development of a bi-modal imaging system, where AFI is used to detect large suspicious regions at which time NBI is used to further interrogate the suspicious lesions based on neovascular patterns. This bi-modal system effectively reduced the false positive rate from 40% to 10%.

The Olympus Corporation has taken this technology one step further and produced a tri-modal system. This multiplatform technology combines high-resolution endoscopy (HRE), AFI, NBI, and magnification into one prototype endoscope. Curvers et al. demonstrated that in a multicenter trial the addition of AFI to HRE increased the detection of both the number of patients and number of lesions with early neoplasia in patient's with BE (Curvers, Singh et al. 2008). In addition the false positive rate of AFI was reduced after detailed inspection with NBI zoom magnification from 81% to 25%.

The use of autofluorescence for polyp detection in the colon was originally studied using a fiberoptic-based system called Onco-LIFE® (Xillix Corporation, Richmond, British Columbia) in 63 patients. Segmental examination during withdrawal was carried out using WLE followed by fluorescent light colonoscopy. Lesions missed at white light examination were recorded as false negatives. White light colonoscopy detected 101 positive lesions and 18 additional adenomas were identified only by fluorescence. The addition of fluorescence imaging resulted in an increase in the adenoma detection rate of 17.8%, which included diminutive and flat adenomatous lesions. The accuracy of endoscopic diagnosis for dysplastic lesions improved from 58% to 72% and for hyperplastic lesions from 62% to 73%. In the noninflamed colon there were no false negatives. False positive lesions were seen with acute and chronic active colitis (Zanati et al. 2005).

In contrast to the number of studies of AFI in Barrett's esophagus, there are no published studies on AFI in patients undergoing surveillance for ulcerative colitis. However, one publication is available of AFI video endoscopy in the colon. Matsuda et al. carried out a pilot study to determine if the newly developed AFI system detected more colorectal polyps than WLE (Matsuda et al. 2008). A modified back-to-back colonoscopy of the right-sided colon was conducted in 167 patients. Subsequent to cecal intubation, patients were randomized to either AFI followed by WL or WL followed by AFI. The total numbers of polyps detected were 100 and 73, respectively. The miss rate for all polyps with AFI (30%) was significantly less than with WL (49%) ($p = 0.01$). Of the neoplastic lesions, at first inspection, 71% were recognized first by the AFI withdrawal technique in contrast to 53% recognized at the first WL withdrawal technique. This is in concordance with the previous fiber optic studies in that the authors concluded that AFI detects more diminutive adenomatous polyps as compared to WL colonoscopy.

The known historical disadvantage of autofluorescence is the high false-positive ratio related to background inflammation in patients with Barrett's and chronic active ulcerative colitis as well as limited levels of fluorescent signal in capacious organs such as the stomach and right colon. Although NBI with magnification has improved the differentiation and reduced the false-positive rate, this limitation may become a major impediment to the widespread acceptance of AFI in clinical practice. Whether a multimodal system will result in a better yield than the traditional Seattle protocol in Barrett's esophagus or the recommended biopsies for ulcerative colitis, remains the goal of future multicenter trials.

13.5.5 MAGNIFYING AND MICROSCOPIC ENDOSCOPY

Magnifying endoscopy provides better visualization of mucosal detail not provided by standard endoscopy and emphasizes mucosal patterns, epithelial crests, and small vessels. High magnification endoscopes use a moveable lens, controlled by the endoscopist, to vary the degree of magnification, which can be up to 120-fold. However, this resolution comes at the price of a limited field of view (Kumagai, Iida, and Yamazaki 2006). To obtain optimal images, which are in focus, a distal attachment is used to stabilize the tissue and maintain a constant distance from the mucosal surface. Most current standard endoscopes have at least an electronic zoom function, which is usually limited to 2× magnification. It is important to note that these standard endoscopes only enlarge the field of view but do not provide microvascular detail. High magnification endoscopy is best carried out in conjunction with either dye spraying or NBI-like technology.

Unlike other digital modalities where classification systems lack agreement and validation, magnifying endoscopy has achieved widespread clinical acceptance for differentiation of sessile lesion in the colon as the surface microstructure of colonic mucosa displays different types of pit patterns; and specific arrangements of colonic crypt orifices. Kudo et al. (2001) have developed a classification system, where the various pit patterns have been divided into five distinct types found in Table 13.1. These features, as described by Kudo, have been the basis for the Paris and more recently the Kyoto consensus of multidisciplinary experts (Kyoto, Japan, 2008) of morphological changes in early gastrointestinal neoplastic lesions (Lambert 2005; Kudo et al. 2008).

TABLE 13.1
Pit Pattern Classification in Colorectal Mucosa

Classification	Description
Type I	Normal mucosa pits are round, regular in size and arrangement.
Type II	Non-neoplastic, hyperplastic polyps, which are larger than the normal pits and are star shaped or onion-like, but are regularly arranged.
Type III$_L$	"L" stands for long or large, seen in polypoid adenomas, where the pits are elongated
Type III$_S$	"S" stands for short or small lesions with compactly arranged pits, which are smaller than normal pits and are characteristically depressed. These tend to be early signs of cancer.
Type IV	Pits of polypoid adenomas that are branched or have a gyrus-like pattern.
Type V	Very irregular pit pattern indicating high-grade dysplasia. This pattern is divided into V_I ("I" for irregular) and V_N ("N" for nonstructural). V_I has pits irregular in shape, size and arrangement and V_N shows an absence of pit pattern.

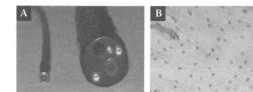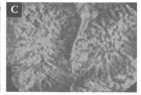

FIGURE 13.6 *In vivo* pathological assessment of Barrett's Esophagus with endocytoscopy system. A: XEC*300 (Olympus Corporation, Tokyo, Japan) endo-cytoscopy probe next to therapeutic gastroscope. B: Normal esophageal squamous mucosa. C: Barrett's mucosa with high-grade dysplasia displaying distorted and complex crypt architecture, high cell density and nuclear atypia (450× magnification, 300 µm × 300 µm field of view, 1% methylene blue).

For instance, the observation of a type 5 change correlates with the risk of submucosal invasion. Similar classifications using high magnification definition are evolving for Barrett's dysplasia and early dysplastic lesions in the stomach.

Endocytoscopy is a prototype light microscopy technique that produces ultra-high magnified images of the superficial mucosa. After the application of 0.5–1% methylene blue, nuclei of the uppermost cellular layer can be visualized. There are currently two prototype probes, one with a magnification of 450 times and a field of view of ~300 µm × 300 µm, while the other has a magnification of 1100 times and a field of view of ~120 µm × 120 µm. The diameter of these probes are 3.5 mm and therefore require a working channel in the therapeutic endoscope of at least 3.7 mm (Inoue, Kudo, and Shiokawa 2005). Figure 13.6 demonstrates the differences between real-time histological images of normal esophageal squamous mucosa and Barrett's mucosa with high-grade dysplasia obtained by *in vivo* endocytoscopy.

Inoue et al. studied esophageal lesions in 28 patients using a prototype endo-cytoscopy integrated endoscope (Inoue et al. 2006). They reported an overall accuracy of endocytoscopy in differentiating between malignant and nonmalignant pathology as 82%. The positive predictive value for malignancy was 94%. The authors also proposed a classification system, but further work and multicenter trials would be necessary to validate these results and to define its role in clinical practice. Limitations of endocytoscopy include the facts that technology is still in a prototype phase, instruments are fragile devices and are hindered by peristalsis, respiratory, and vascular motion, as well as gastric secretions as they require direct contact with the mucosa.

13.6   CONFOCAL LASER ENDOMICROSCOPY

Confocal laser endomicroscopy (CLE) provides *in vivo* histology during endoscopy with a depth of 100–200 µm using a laser wavelength of ~488 nm. The resultant field of view is ~500 × 500 µm at ~7 µm tissue slices. The endoscope incorporates a miniature confocal microscope in the distal tip of a conventional video endoscope allowing simultaneous white light and confocal microscopy. For contrast enhancement, intravenous fluorescein is injected and distributes throughout the entire mucosa leading to *in vivo* cross-sectional visualization of mucosal and subsurface cellular morphology.

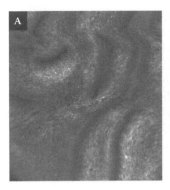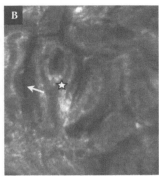

FIGURE 13.7 *In vivo* CLE images of BE. (A) Roundish appearance of mucin-filled goblet cells in a patient with low-grade dysplasia. (B) Glandular structure is noticeable more disorganized, and the basement membrane is irregular (arrow) in a patient with high-grade dysplasia. An irregular capillary is also detected, as capillary leakage of fluorescein leads to an increase in signal intensity (star). (Courtesy of Sharmila Anandasabapathy, M.D., Mount Sinai School of Medicine, New York.)

A CLE probe has also been developed and can now fit into the working channel of most gastrointestinal (GI) endoscopes (Meining et al. 2007). However, this miniaturization comes at the cost of decreased image resolution with tissue slices of ~15 µm. This probe-based technology is currently being evaluated in clinical trials of Barrett's esophagus (see Figure 13.7), ulcerative colitis, and cholangiocarcinoma. A novel application of this technology involves the study of molecular probes in form of fluorescent peptides that bind to dysplastic tissue in the esophagus and colon that enhance endoscopic detection of early lesions (Hsiung et al. 2008).

Current recommendations for Barrett's surveillance include careful examination, biopsies of suspicious areas, and then a systematic biopsy protocol involving 4 quadrant random biopsies every one to two cm for the length of the Barrett's. Despite this vigorous protocol, with multiple biopsies, the diagnostic yield for dysplasia during surveillance has been highly variable as detection of dysplasia can be difficult. Kiesslich et al. have applied CLE to Barrett's esophagus to establish a endomicrospy classification system, the confocal Barrett's classification, which incorporates vascular structure and cell patterns to distinguish between gastric mucosa, Barrett's epithelium, and dysplasia (Kiesslich et al. 2006). Mucosal pathology obtained at biopsy was used as the reference standard. They demonstrated that the classification system had a sensitivity of 92.9%, specificity of 98.4%, and accuracy of 97.4% for predicting Barrett's associated neoplasia (Kiesslich et al. 2006). This landmark study was recently confirmed by Dunbar et al., where CLE-guided biopsies improved the yield for neoplastic detection from 17% to 33% in patients with suspected inapparent HGD (Dunbar et al. 2009). They also demonstrated that CLE-targeted biopsies reduced the number of biopsies and that some patients without neoplasia were able to completely forego mucosal biopsies.

CLE has also been used to image neoplasia in the colon. Neoplastic changes could be predicted in polyps with a sensitivity, specificity, and accuracy rate of 97.4%, 99.4%, and 99.2%, respectively (Kiesslich et al. 2004). Using a probe system,

Buchner et al. conducted a prospective, blinded trial where 103 polyps were imaged with a sensitivity of 80% and specificity of 94% for the detection of adenomatous polyps (Buchner et al. 2008). Although this study implies a potential usefulness for probe-based optical biopsy, it falls short of supplying the endoscopist with sufficient endoscopic confidence as to whether a polyp may be ignored or must be removed.

In longstanding chronic ulcerative colitis there is an increased risk of neoplastic transformation. Surveying the colon affected by chronic inflammatory disease poses a significant challenge because of the capacity of the colon, the difficulty of obtaining an adequate colon prep and that the dysplastic areas may be focal and occult. For example, standard biopsy forceps jaws are 2 × 8 mm with a surface area 16 mm^2, and the average colon length under observation is 1000 mm long with a diameter of 100 mm. This presents to the endoscopist a surface area of more than 300,000 mm^2 that requires careful observation. Using the usual biopsy protocol means that <0.2% of the surface area is sampled (Wallace 2009). Therefore, the probability of finding diseased tissue can indeed be a hunt for a needle in the haystack. This is supported in practice by the fact that, traditionally, at least 32 biopsies are recommended for surveillance.

In a multicenter trial of patients with chronic ulcerative colitis, the combination of chromoendoscopy using methylene blue and confocal endoscopy was carried out.

The chromoendoscopy acted as a "red flag" to highlight potentially dysplastic areas. The addition of CLE resulted in the detection of 4.7 times more dysplastic lesions than conventional colonoscopy (Kiesslich et al. 2007).

Although CLE has been primarily used for investigation of neoplastic lesions (Kiesslich et al. 2004), its capacity for real-time histological evaluation has been studied in a variety of nondysplastic conditions such as collagenous colitis (Zambelli et al. 2008), celiac disease (Leong et al. 2008), and graft versus host disease (Bojarski et al. 2009). It has the exciting potential of eliciting, in a real-time manner, the pathophysiology of gastrointestinal function and disease. The discovery of cell shedding and epithelial gaps (Watson et al. 2005) is an example that could demonstrate a link between transient structural defects, bacteria, and bacterial translocations as a possible initiator in inflammatory bowel disease (Kiesslich et al. 2007).

CLE is a relatively new technology and has the potential to become a clinically relevant imaging adjunct during endoscopic examinations. Under ideal conditions, it provides *in vivo* near histological resolution, which can be useful to guide biopsies and increase diagnostic yield. However, additional technological improvements are required. For example, different imaging wavelengths to increase penetration depth; 3D imaging to obtain multiple focal zones; wider field of view to increase the efficiency of screening; and improved imagery storage/retrieval/display to allow comprehensive examination of the GI tract. Combination of fluorescence markers and molecular beacons could potentially further increase the sensitivity and specificity of CLE, although the added complexity and trade-off remain an active research area.

13.7 OPTICAL COHERENCE TOMOGRAPHY

Time-domain optical coherence tomography (TD-OCT; Huang et al. 1991) was developed as an imaging modality to visualize subsurface tissue architecture at a resolution (~10 μm in tissue) approaching histology. Continuous improvements have resulted in an imaging technique that demonstrates great promise as a noninvasive optical biopsy tool. OCT is considered to be analogous to ultrasound, however, instead of measuring backscattered pressure waves, OCT measures backscattered photons using an interferometer, avoiding the difficult task of directly measuring the time-of-flight of the backscattered light. Advantages of OCT overcompeting modalities, such as endoscopic ultrasound, include a 10× improvement in resolution, without the need for acoustic coupling media (i.e., water). Additionally, OCT provides high resolution imaging to a depth of ~1–2 mm, several times deeper than that of confocal laser endoscopy ~150–200 μm (Yang, Tang et al. 2005). For imaging targets that are not stationary, such as red blood cells, blood flow detection functionality has been incorporated into OCT. Doppler optical coherence tomography (DOCT) has been used to investigate fluid dynamics, interstitial blood flow, retinal blood flow, and hereditary hemorrhagic telangiectasia (Moger et al. 2004; Standish et al. 2007; White et al. 2003; Tang et al. 2003).

Preliminary work by Yang et al. have used endoscopic (E) DOCT to document variations in blood vessel density and location in several GI pathologies (Yang, Tang et al. 2005) such as Barrett's esophagus, esophageal varices, gastric antral vascular ectasia, and portal hypertensive gastrophy (see Figures 13.8–13.9). Vascular morphology is an important criterion of dysplasia that has been well describe by Yao et al. and the Kyoto consensus of multidisciplinary experts using conventional high magnification (×100) endoscopy (Yao et al. 2007; Kudo et al. 2008). The applica-

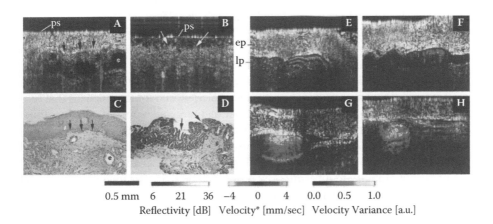

0.5 mm   6   21   36   −4   0   4   0.0   0.5   1.0
Reflectivity [dB] Velocity* [mm/sec] Velocity Variance [a.u.]

FIGURE 13.8 (See color insert) EDOCT images of Barrett's esophagus and esophageal varices from four different patients. (A,C) Subsquamous BE with mucosal glands (asterisks) underneath the clearly delineated epithelial-lamina propria interface (arrows). (B,D) Barrett's esophagus with superficial glandular structure (arrows) and microvasculature close to the surface. (E,F) Doppler images of blood flow in the dilated variceal vessels. The blood flow velocity in (E) is more than three times greater than that seen in normal esophagus. (G,H) Velocity-variance images of microcirculation in dilated variceal vessels. (C,D) CD34, Orig. mag. ×10. PS, External surface of the imaging tip. (Reprinted from Yang, V. X. D., Tang, M. L. et al.. 2005. *Gastrointestinal Endoscopy* 61 (7): 879–890. With permission from Elsevier.)

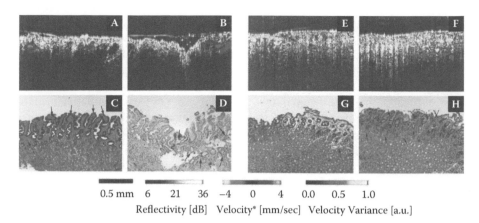

0.5 mm   6   21   36   −4   0   4   0.0   0.5   1.0
Reflectivity [dB]   Velocity* [mm/sec]   Velocity Variance [a.u.]

FIGURE 13.9 (See color insert)   (A,B) Color EDOCT images of GAVE in two patients. (C) Consistent with the H&E staining, dilated microvasculature (arrows) is present immediately beneath the tissue surface. (D) Consistent with the CD34 staining, dilated microvasculature (arrows) is present immediately beneath the tissue surface. (E) Color EDOCT images of portal hypertensive gastropathy showing superficial vessels, consistent with the H&E histology in (G) and (H) marked by arrows. (F) Velocity variance EDOCT images of PHG, showing superficial vessels, consistent with the H&E histology in (G) and (H) marked by arrows. (C,D,G,H) Orig. mag. ×10. (Reprinted from Yang, V. X. D., Tang, M. L. et al., 2005. *Gastrointestinal Endoscopy* 61 (7): 879–890. With permission from Elsevier.)

tion of DOCT presents an opportunity for potential investigation in the detection of neovascularization, which is associated with early dysplasia.

Although the clinical extension of OCT to image the luminal surface of the GI tract has been feasible since 1997 (Tearney et al. 1997), very few studies have been reported demonstrating the clinical utility of OCT in the GI tract. TD-OCT had several imaging limitations due to the achievable scanning rates associated with the system hardware. This resulted in a small field of view and imaging speeds that limited its usefulness during routine surveillance and targeted biopsy. To address these problems there has recently been a platform shift to a second-generation OCT system called Fourier domain (FD) OCT. This new platform benefits from an improved signal-to-noise ratio (as light from all depths contribute to the signal), improved phase stability (ability to measure slow moving targets), higher imaging speeds and reduced system hardware and complexity (Choma et al. 2003). Current state-of-the-art ultra-high resolution FD-OCT systems have achieved sub-micron spatial resolutions *ex-vivo* (Bizheva et al. 2005), allowing for the detection of subcellular features along with an ability to detect blood flow down to the capillary bed (Mariampillai et al. 2008). These excellent structural and blood flow detection imaging characteristics suggest OCT will play an important role in accurate noninvasive optical biopsy in the GI tract.

Initial studies have been performed to determine the ability of FD-OCT to diagnose specific pathologies, such as specialized intestinal metaplasia (SIM). Evans et al. have outlined an algorithm for the detection of SIM using OCT, where two blinded readers were 81% and 81% sensitive, and 66% and 57% specific for diagnosis

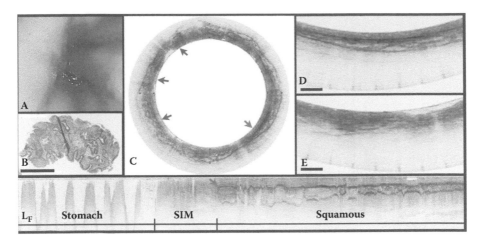

FIGURE 13.10 Barrett's esophagus. (A) Videoendoscopic image demonstrates an irregular SCJ. (B) Histopathologic image of a biopsy specimen obtained from the SCJ demonstrates SIM without dysplasia (H&E, orig. mag. ×2). (C) Cross-sectional OFDI image reveals both the normal layered appearance of squamous mucosa (red arrow, expanded in [D]) and tissue that satisfies the OCT criteria for SIM (blue arrows, expanded in [E]). F, Longitudinal section across the gastroesophageal junction shows the transition from squamous mucosa to SIM to cardia. The length of the BE segment is 7 mm in this OFDI reconstruction. Scale bars and tick marks represent 1 mm. (Reprinted from Suter, M. et al. 2008. *Gastrointestinal Endoscopy* 68 (4): 745–753. With permission from Elsevier.)

of SIM at the squamocolumnar junction (SCJ) (Evans et al. 2007). This research, and previous OCT feasibility studies (Poneros et al. 2001; Isenberg et al. 2005), were very promising but were limited due to the sampling nature of the image acquisition, ultimately limiting the usefulness of this technology to that of a random biopsy. With the advent of high-speed FD-OCT (Oh et al. 2005; Adler, Huber, and Fujimoto 2007) and recently developed balloon-centering optical catheters (Vakoc et al. 2007), large fields of view can now be imaged during routine GI imaging procedures. Suter et al. have recently published a feasibility study of large area upper GI comprehensive optical microscopy via FD-OCT (Suter et al. 2008). The entire distal esophagus (~6.0 cm) was imaged with a resolution of 20 µm × 8 µm × 50 µm (radial, axial, z-step) allowing for the visualization of microscopic features consistent with histological findings such as squamous mucosa, cardia, SIM with/without dysplasia, and esophageal erosion. Figure 13.10 is an example data set from a patient with a history of Barrett's esophagus, where the benefit of using OCT to scan the entire region of interest is demonstrated. This technique is not limited to imaging only the upper GI tract, as similar techniques are relevant for imaging the lower GI tract.

Adler et al. have demonstrated the potential of FD-OCT to distinguish structural features of normal and colorectal tissue in addition to monitoring the healing process post-radiation treatment for proctitis (Adler, Zhou et al. 2009). Beyond structural tissue classification, recent advances in signal processing techniques have improved the ability to detect almost the complete vascular network within an acquired three-dimensional (3D) FD-OCT image volume (Mariampillai et al. 2008). Figure 13.11

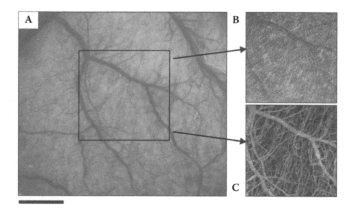

FIGURE 13.11   Dorsal skin-fold window model. A) White light microscopy of window, demonstrating the visible vasculature. The black box represents the approximate location of the OCT region of interest. B) Structural projection of 3D-OCT image volume. C) Speckle variance projection of 3D-OCT image volume, highlighting the vascular network. (Courtesy of Michael Leung, Princess Margaret Hospital, Toronto, Canada.)

demonstrates this blood detection technique, termed speckle variance, in a dorsal skin-fold window chamber animal model, where the exceptional vessel detection is possible without exogenous contrast agents. This additional contrast mechanism may extend the ability of OCT to qualify and quantify different pathologies in the GI tract.

Essential requirements of OCT, as a realistic optical biopsy tool, necessitates a high negative predict value that justifies not taking physical biopsies. The representation of tissue will require enough histological detail to have the confidence of targeting and interrogating suspicious areas for biopsy, saving time and money associated with pathological review. It seems there is a degree of optimism, that in the foreseeable future, OCT can accomplish this task and will have a major impact on surveillance of micro intestinal pathologies. However, several hurdles exist before this technique becomes a standard imaging option for gastroenterologists. Although the high resolution of OCT is very useful for identifying microstructural architecture, there are current issues with the manipulation, evaluation, and storage of the associated data. A typical pullback data set can approach 40 GB and consist of ~1200 individual cross-sectional images (Suter et al. 2008). It may prove difficult for acceptance of this approach, as it would take an extensive amount of time and effort by highly qualified personnel, to thoroughly analyze each of the large and complex data sets. Alternatively, large regions of interest could be scanned, where abnormal images or regions are highlighted, via pattern recognition algorithms, to filter the data prior to clinical review.

As the FD-OCT platform matures, and an imaging protocol is developed for use in endoscopic procedures such as non-invasive optical biopsy, there exists an additional benefit of guided focal disease treatment. For example, as a FD-OCT image is acquired, the physical location of the probe could be co-localized to the acquired image. If the suspicious region requires treatment, the endoscopist could select the area for ablation and the FD-OCT probe would be automatically repositioned to treat the area through an additional optical channel dedicated to laser-ablation of the tissue. These potential applications outline the usefulness of OCT as a tool to be used by gastroenterologists, are technically feasible in the near future and maybe become the standard for non-invasive optical imaging in the GI tract.

13.8 CONCLUSIONS

In the early seventies when fiberoptic endoscopes were being introduced into clinical practice, the emphasis was primarily on assessing patients with symptoms or on the diagnosis and biopsies of abnormalities first seen with barium contrast studies. In the intervening decades to date, there has been remarkable progress in endoscope design driven by clinical need and technological advances that have resulted in brighter illumination of tissue surface, mucosal contrast, and the capture and transmission of high-resolution images. The endoscopist now has an extensive toolset that provides clear and robust visualization of the mucosa, hopefully leading to better patient care. This improved imaging, coupled with endoscopic maturity, has altered the aim of endoscopy to focus on the diagnosis of dysplasia and pre-invasive pathologies (early cancer). This gradual evolution has been led by Japanese endoscopits. In Japan, at the beginning of this era, cancer of the stomach was singled out as a major public health initiative and led to widespread screening of asymptomatic citizens. This coupled with the fact that the major endoscope manufacturers reside in Japan, has resulted in a concerted effort to improve instrumentation to facilitate the diagnosis and characterization of mucosal abnormalities and a drive to detect smaller and smaller lesions. Originally, it was felt by Western authorities that these early lesions in the stomach and even in the colon were unique to the Japanese population. However, it has subsequently been shown that the same lesions and evolution from dysplasia to invasive and metastatic cancer are somewhat similar in Western populations. The assumption has been that Western endoscopists were perhaps not as observant or as careful as their Japanese counterparts. Rene Lambert, in his 2004 editorial, implied that Western endoscopists were looking at the mucosa but were in fact not seeing the mucosa. He termed this concept "eyes wide shut" (Lambert, Jeannerod, and Rey 2004). The identification of large lesions is ordinarily no longer a challenge, but with increasing emphasis on endoscopic therapy for the cure of mucosal cancer, the search continues to improve our ability to reach a high threshold of endoscopic detection of "occult" lesions.

One of the early phases of improved mucosal detection has revolved around the use of various dye spraying contrast agents. These have been particularly useful in squamous esophagus, stomach, and colon. They have facilitated improved detection using "ordinary" white light endoscopy to identify abnormal areas and to delineate the borders between dysplastic and uninvolved mucosa. This ability to identify field defects, function as a red flag or waving hand to direct the interest of the endoscopist to these areas and to facilitate targeted biopsies. This method of abnormal mucosal detection has evolved quite rapidly due to the development of chip-based technologies associated with high definition, high resolution, higher pixel densities, and endoscopic miniaturization.

In an attempt to improve wide field discrimination, and to avoid the use of dye spraying and its associated false-positive rate, a number of image enhancing technologies have been introduced in the last 10 years. These include: NBI, FICE, i-scan, autofluorescence, and magnification. These new technologies have thus far yielded conflicting results and have had a variable impact on the improved diagnosis of dysplasia. It seems that studies from leading experts have resulted in confounding findings. The conclusions were that the image-enhancing technologies are useful, but perhaps the most important aspect to the technology is the quantity and quality of time spent to observe the region of interest. We suspect the final solution for identification of field defects will be a combination of the well-trained eye in conjunction with these new technologies either alone or in multimodal platforms. With these suspicious regions highlighted, the dream of every endoscopist would be to have real time H&E equivalent histology of the mucosal layer. This is what we term "optical biopsy." One step towards the achievement of this optical biopsy dream involves using light to noninvasively interrogate subsurface structures.

Currently, the leading technology in this area is confocal endomicroscopy. There are still some mechanical and technical aspects of the instrumentation, which require improvement before widespread clinical adoption of this technology can occur. Another promising technology, OCT, has additional features, such as increased depth of imaging and wide field of view, but considerable research and engineering hurdles need to be overcome before this platform will be available for clinical use. These new optical biopsy modalities will shift histological diagnosis from the tissue extraction-based pathologist to the real-time optical biopsy endoscopist. This will take some time, but will be inevitable. A major shift in education and training of the endoscopist will be required before these new concepts and diagnostic algorithms are incorporated into daily practice. We speculate these technological advancements will result in evidence-based improved pre-cancerous detection yielding a reduction in the cancer burden, superior patient care, and cost-effective screening modalities.

REFERENCES

Adler, A., J. Aschenbeck, T. Yenerim, M. Mayr, A. Aminalai, R. Drossel, A. Schrˆder, M. Scheel, B. Wiedenmann, and T. Rˆsch. 2009. Narrow-band versus white-light high definition television endoscopic imaging for screening colonoscopy: A prospective randomized trial. *Gastroenterology* 136 (2): 410–416.

Adler, A., H. Pohl, I. S. Papanikolaou, H. Abou-Rebyeh, G. Schachschal, W. Veltzke-Schlieker, A. C. Khalifa, E. Setka, M. Koch, and B. Wiedenmann. 2008. A prospective randomised study on narrow-band imaging versus conventional colonoscopy for adenoma detection: Does narrow-band imaging induce a learning effect? *British Medical Journal* 57 (1): 59.

Adler, D. C., R. Huber, and J. G. Fujimoto. 2007. Phase-sensitive optical coherence tomography at up to 370,000 lines per second using buffered Fourier domain mode-locked lasers. *Optics Letters* 32 (6): 626–628.

Adler, D. C., C. Zhou, T. H. Tsai, J. Schmitt, Q. Huang, H. Mashimo, and J. G. Fujimoto. 2009. Three-dimensional endomicroscopy of the human colon using optical coherence tomography. *Optics Express* 17 (2): 784–786.

Astorga-Wells, J., S. Vollmer, S. Tryggvason, T. Bergman, and H. Jornvall. 2005. Microfluidic electrocapture for separation of peptides. *Analytical Chemistry* 77 (22): 7131–7136.

Backman, V., M. B. Wallace, L. T. Perelman, J. T. Arendt, R. Gurjar, M. G. MuÃàller, Q. Zhang, G. Zonios, E. Kline, T. McGillican, S. Shapshay, T. Valdez, K. Badizadegan, J. M. Crawford, M. Fitzmaurice, S. Kabani, H. S. Levin, M. Seiler, R. R. Dasari, I. Itzkan, J. Van Dam, and M. S. Feld. 2000. Detection of preinvasive cancer cells. *Nature* 406 (6791): 35–36.

Bakker Schut, T. C., M. J. H. Witjes, H. J. C. M. Sterenborg, O. C. Speelman, J. L. N. Roodenburg, E. T. Marple, H. A. Bruining, and G. J. Puppels. 2000. *In vivo* detection of dysplastic tissue by Raman spectroscopy. *Analytical Chemistry* 72 (24): 6010–6018.

Bizheva, K., A. Unterhuber, B. Hermann, B. Povazay, H. Sattmann, A. F. Fercher, W. Drexler, M. Preusser, H. Budka, A. Stingl, and T. Le. 2005. Imaging *ex vivo* healthy and pathological human brain tissue with ultra-high-resolution optical coherence tomography. *Journal of Biomedical Optics* 10 (1): 11006.

Bojarski, C., U. Gunther, K. Rieger, F. Heller, C. Loddenkemper, M. Grynbaum, L. Uharek, M. Zeitz, and J. C. Hoffmann. 2009. *In vivo* diagnosis of acute intestinal graft-versus-host disease by confocal endomicroscopy. *Endoscopy* 41 (5): 433.

Buchner, A. M., M. S. Ghabril, M. Krishna, H. C. Wolfsen, and M. B. Wallace. 2008. High-resolution confocal endomicroscopy probe system for *in vivo* diagnosis of colorectal neoplasia. *Gastroenterology* 135 (1): 295–295.

Choma, M. A., M. V. Sarunic, C. Yang, and J. A. Izatt. 2003. Sensitivity advantage of swept source and Fourier domain optical coherence tomography. *Optics Express* 11 (18): 2183–2189.

Curvers, W. L., C. J. Bohmer, R. C. Mallant-Hent, A. H. Naber, C. I. J. Ponsioen, K. Ragunath, R. Singh, M. B. Wallace, H. C. Wolfsen, and L. M. Song. 2008. Mucosal morphology in Barrett's esophagus: Interobserver agreement and role of narrow band imaging. *Endoscopy* 40 (10): 799.

Curvers, W. L., R. Singh, L. M. Wong-Kee Song, H. C. Wolfsen, K. Ragunath, K. Wang, M. B. Wallace, and P. Fockens. 2008. Bergman JJGHM. Endoscopic tri-modal imaging for detection of early neoplasia in Barrett's oesophagus: A multi-centre feasibility study using high-resolution endoscopy, autofluorescence imaging and narrow band imaging incorporated in one endoscopy system. *Gut* 57: 167–172.

Dhar, A., K. S. Johnson, M. R. Novelli, S. G. Bown, I. J. Bigio, L. B. Lovat, and S. L. Bloom. 2006. Elastic scattering spectroscopy for the diagnosis of colonic lesions: Initial results of a novel optical biopsy technique. *Gastrointestinal Endoscopy* 63 (2): 257–261.

Dittrich, P. S., K. Tachikawa, and A. Manz. 2006. Micro total analysis systems. Latest advancements and trends. *Analytical Chemistry* 78 (12): 3887–3908.

Dunbar, K. B., P. Okolo Iii, E. Montgomery, and M. I. Canto. 2009. Confocal laser endomicroscopy in Barrett's esophagus and endoscopically inapparent Barrett's neoplasia: A prospective, randomized, double-blind, controlled, crossover trial. *Gastrointestinal Endoscopy* 70 (4): 645–654.

Evans, J. A., B. E. Bouma, J. Bressner, M. Shishkov, G. Y. Lauwers, M. Mino-Kenudson, N. S. Nishioka, and G. J. Tearney. 2007. Identifying intestinal metaplasia at the squamocolumnar junction by using optical coherence tomography. *Gastrointestinal Endoscopy* 65 (1): 50–56.

Ferlay, J., F. Bray, P. Pisani, and D. M. Parkin. 2004. GLOBOCAN 2002: Cancer Incidence, Mortality and Prevalence Worldwide IARC Cancer Base No. 5. version 2.0. Lyon: IARCPress.

Georgakoudi, I., B. C. Jacobson, J. Van Dam, V. Backman, M. B. Wallace, M. G. Muller, Q. Zhang, K. Badizadegan, D. Sun, and G. A. Thomas. 2001. Fluorescence, reflectance, and light-scattering spectroscopy for evaluating dysplasia in patients with Barrett's esophagus. *Gastroenterology* 120 (7): 1620–1629.

Goda, K., H. Tajiri, M. Ikegami, M. Urashima, T. Nakayoshi, and M. Kaise. 2007. Usefulness of magnifying endoscopy with narrow band imaging for the detection of specialized intestinal metaplasia in columnar-lined esophagus and Barrett's adenocarcinoma. *Gastrointestinal Endoscopy* 65 (1): 36–46.

Gono, K., K. Yamazaki, N. Doguchi, T. Nonami, T. Obi, M. Yamaguchi, N. Ohyama, H. Machida, Y. Sano, S. Yoshida, Y. Hamamoto, and T. Endo. 2003. Endoscopic observation of tissue by narrowband illumination. *Optical Review* 10 (4): 211–215.

Guelrud, M., and I. Herrera. 1998. Acetic acid improves identification of remnant islands of Barrett's epithelium after endoscopic therapy. *Gastrointestinal Endoscopy* 47 (6): 512–515.

Hoffman, A., R. Kiesslich, A. Bender, M. F. Neurath, B. Nafe, G. Herrmann, and M. Jung. 2006. Acetic acidñguided biopsies after magnifying endoscopy compared with random biopsies in the detection of Barrett's esophagus: A prospective randomized trial with crossover design. *Gastrointestinal Endoscopy* 64 (1): 1–8.

Hsiung, P. L., J. Hardy, S. Friedland, R. Soetikno, C. B. Du, A. P. Wu, P. Sahbaie, J. M. Crawford, A. W. Lowe, and C. H. Contag. 2008. Detection of colonic dysplasia *in vivo* using a targeted heptapeptide and confocal microendoscopy. *Nature* 14: 454–458.

Huang, D., E. A. Swanson, C. P. Lin, J. S. Schuman, W. G. Stinson, W. Chang, M. R. Hee, T. Flotte, K. Gregory, C. A. Puliafito, and J. G. Fujimoto. 1991. Optical coherence tomography. *Science* 254 (5035): 1178–1181.

Inoue, H, S Kudo, and A Shiokawa. 2005. Technology insight: Laser-scanning confocal microscopy and endocytoscopy for cellular observation of the gastrointestinal tract. *Nature Clinical Practice Gastroenterology and Hepatology* 2 (1): 31–37.

Inoue, H., K. Sasajima, M. Kaga, S. Sugaya, Y. Sato, Y. Wada, M. Inul, H. Satodate, S. E. Kudo, and S. Kimura. 2006. Endoscopic *in vivo* evaluation of tissue atypia in the esophagus using a newly designed integrated endocytoscope: A pilot trial. *Endoscopy* 38 (9): 891–895.

Isenberg, G., M. V. Sivak Jr, A. Chak, R. C. K. Wong, J. E. Willis, B. Wolf, D. Y. Rowland, A. Das, and A. Rollins. 2005. Accuracy of endoscopic optical coherence tomography in the detection of dysplasia in Barrett's esophagus: A prospective, double-blinded study. *Gastrointestinal Endoscopy* 62 (6): 825–831.

Izuishi, K., H. Tajiri, T. Fujii, N. Boku, A. Ohtsu, T. Ohnishi, M. Ryu, T. Kinoshita, and S. Yoshida. 1999. The histological basis of detection of adenoma and cancer in the colon by autofluorescence endoscopic imaging. *Endoscopy* 31 (7): 511–516.

Jung, M, and R Kiesslich. 1999. Chromoendoscopy and intravital staining techniques. *BailliÈre's best practice & research. Clinical Gastroenterology* 13 (1): 11.

Kara, M. A., M. Ennahachi, P. Fockens, F. J. W. ten Kate, and J. J. Bergman. 2006. Detection and classification of the mucosal and vascular patterns (mucosal morphology) in Barrett's esophagus by using narrow band imaging. *Gastrointestinal Endoscopy* 64 (2): 155–166.

Kara, M. A., F. P. Peters, P. Fockens, F. J. W. ten Kate, and J. J. Bergman. 2006. Endoscopic video-autofluorescence imaging followed by narrow band imaging for detecting early neoplasia in Barrett's esophagus. *Gastrointestinal Endoscopy* 64 (2): 176–185.

Kawahara, Y., R. Takenaka, H. Okada, S. Kawano, M. Inoue, T. Tsuzuki, D. Tanioka, K. Hori, and K. Yamamoto. 2009. Novel chromoendoscopic method using an acetic acid-indiocarmine mixture for diagnostic accuracy in delineating the margin of early gastic cancers. *Digestive Endoscopy* 21 (1): 14.

Kendall, C., N. Stone, N. Shepherd, K. Geboes, B. Warren, R. Bennett, and H. Barr. 2003. Raman spectroscopy, a potential tool for the objective identification and classification of neoplasia in Barrett's oesophagus. *Journal of Pathology* 200 (5): 602–609.

Kiesslich, R., M. Goetz, E. M. Angus, Q. Hu, Y. Guan, C. Potten, T. Allen, M. F. Neurath, N. F. Shroyer, and M. H. Montrose. 2007. Identification of epithelial gaps in human small and large intestine by confocal endomicroscopy. *Gastroenterology* 133 (6): 1769–1778.

Kiesslich, R., L. Gossner, M. Goetz, A. Dahlmann, M. Vieth, M. Stolte, A. Hoffman, M. Jung, B. Nafe, and P. R. Galle. 2006. *In vivo* histology of Barrett's esophagus and associated neoplasia by confocal laser endomicroscopy. *Clinical Gastroenterology and Hepatology* 4 (8): 979–987.

Kiesslich, R., J. Burg, M. Vieth, J. Gnaendiger, M. Enders, P. Delaney, A. Polglase, W. McLaren, D. Janell, S. Thomas, B. Nafe, P. R. Galle, and M. F. Neurath. 2004. Confocal laser endoscopy for diagnosing intraepithelial neoplasias and colorectal cancer *in vivo*. *Gastroenterology* 127 (3): 706–713.

Kondo, H., H. Fukuda, H. Ono, T. Gotoda, D. Saito, K. Takahiro, K. Shirao, H. Yamaguchi, and S. Yoshida. 2001. Sodium thiosulfate solution spray for relief of irritation caused by Lugol's stain in chromoendoscopy. *Gastrointestinal Endoscopy* 53 (2): 199–202.

Kudo, S., R. Lambert, J. I. Allen, H. Fujii, T. Fujii, H. Kashida, T. Matsuda, M. Mori, H. Saito, and T. Shimoda. 2008. Nonpolypoid neoplastic lesions of the colorectal mucosa. *Gastrointestinal Endoscopy* 68 (4 Suppl): S3.

Kudo, S., C. A. Rubio, C. R. Teixeira, H. Kashida, and E. Kogure. 2001. Pit pattern in colorectal neoplasia: Endoscopic magnifying view. *Endoscopy* 33 (4): 367–373.

Kumagai, Y., M. Iida, and S. Yamazaki. 2006. Magnifying endoscopic observation of the upper gastrointestinal tract. *Digestive Endoscopy* 18 (3): 165.

Kwon, R. S., D. G. Adler, B. Chand, J. D. Conway, D. L. Diehl, S. V. Kantsevoy, P. Mamula, S. A. Rodriguez, R. J. Shah, and L. M. Wong Kee Song. 2009. High-resolution and high-magnification endoscopes. *Gastrointestinal Endoscopy* 69 (3): 399–407.

Lambert, R. 2005. Endoscopic classification review group: Update on the Paris classification of superficial neoplastic lesions in the digestive tract. *Endoscopy* 37: 570–578.

Lambert, R., M. Jeannerod, and J. F. Rey. 2004. Eyes wide shut. *Endoscopy* 36 (8): 723–725.

Lee, Y. C., C. P. Wang, C. C. Chen, H. M. Chiu, J. Y. Ko, P. J. Lou, T. L. Yang, H. Y. Huang, M. S. Wu, and J. T. Lin. 2009. Transnasal endoscopy with narrow-band imaging and Lugol staining to screen patients with head and neck cancer whose condition limits oral intubation with standard endoscope (with video). *Gastrointestinal Endoscopy* 69 (3): 408–417.

Leong, R. W. L., N. Q. Nguyen, C. G. Meredith, S. AlñSohaily, D. Kukic, P. M. Delaney, E. R. Murr, J. Yong, N. D. Merrett, and A. V. Bjankin. 2008. *in vivo* confocal endomicroscopy in the diagnosis and evaluation of celiac disease. *Gastroenterology* 135 (6): 1870–1876.

Lovat, L. B., K. Johnson, G. D. Mackenzie, B. R. Clark, M. R. Novelli, S. Davies, M. O'Donovan, C. Selvasekar, S. M. Thorpe, D. Pickard, R. Fitzgerald, T. Fearn, I. Bigio, and S. G. Bown. 2006. Elastic scattering spectroscopy accurately detects high grade dysplasia and cancer in Barrett's oesophagus. *Gut* 55 (8): 1078–1083.

Mariampillai, A., B. A. Standish, E. H. Moriyama, M. Khurana, N. R. Munce, M. K. K. Leung, J. Jiang, A. Cable, B. C. Wilson, I. A. Vitkin, and V. X. D. Yang. 2008. Speckle variance detection of microvasculature using swept-source optical coherence tomography. *Optics Letters* 33 (13): 1530–1532.

Matsuda, T., Y. Saito, K. I. Fu, T. Uraoka, N. Kobayashi, T. Nakajima, H. Ikehara, Y. Mashimo, T. Shimoda, and Y. Murakami. 2008. Does autofluorescence imaging videoendoscopy system improve the colonoscopic polyp detection rate?—A pilot study. *American Journal of Gastroenterology* 103 (8): 1926–1932.

Meining, A., D. Saur, M. Bajbouj, V. Becker, E. Peltier, H. HoÄäfler, C. H. von Weyhern, R. M. Schmid, and C. Prinz. 2007. *In vivo* histopathology for detection of gastrointestinal neoplasia with a portable, confocal miniprobe: An examiner blinded analysis. *Clinical Gastroenterology and Hepatology* 5 (11): 1261–1267.

Moger, J., S. J. Matcher, C. P. Winlove, and A. Shore. 2004. Measuring red blood cell flow dynamics in a glass capillary using Doppler optical coherence tomography and Doppler amplitude optical coherence tomography. *Journal of Biomedical Optics* 9 (5): 982–994.

Molckovsky, A., L. M. Wong Kee Song, M. G. Shim, N. E. Marcon, and B. C. Wilson. 2003. Diagnostic potential of near-infrared Raman spectroscopy in the colon: Differentiating adenomatous from hyperplastic polyps. *Gastrointestinal Endoscopy* 57 (3): 396–402.

Nakayoshi, T., H. Tajiri, K. Matsuda, M. Kaise, M. Ikegami, and H. Sasaki. 2004. Magnifying endoscopy combined with narrow band imaging system for early gastric cancer: Correlation of vascular pattern with histopathology (including video). *Endoscopy* 36 (12): 1080–1084.

Oh, W. Y., S. H. Yun, G. J. Tearney, and B. E. Bouma. 2005. 115 kHz tuning repetition rate ultra-high-speed wavelength-swept semiconductor laser. *Optics Letters* 30 (23): 3159–3161.

Osawa, H., M. Yoshizawa, H. Yamamoto, H. Kita, K. Satoh, H. Ohnishi, H. Nakano, M. Wada, M. Arashiro, and M. Tsukui. 2008. Optimal band imaging system can facilitate detection of changes in depressed-type early gastric cancer. *Gastrointestinal Endoscopy* 67 (2): 226–234.

Paggi, S., F. Radaelli, A. Amato, G. Meucci, G. Mandelli, G. Imperiali, G. Spinzi, N. Terreni, N. Lenoci, and V. Terruzzi. 2009. The impact of narrow band imaging in screening colonoscopy: A randomized controlled trial. *Clinical Gastroenterology and Hepatology* 7 (10): 1049–1054.

Peitz, U., and P. Malfertheiner. 2002. Chromoendoscopy: From a research tool to clinical progress. *Digestive Diseases* 20: 111–119.

Ponchon, T., M. G. Lapalus, J. C. Saurin, C. Robles-Medranda, C. Chemaly, B. Parmentier, and O. Guillaud. 2007. Could narrow band imaging (NBI) replace lugol staining for the detection of esophageal squamous cell carcinoma? *Gastrointestinal Endoscopy* 65 (5): AB343.

Poneros, J. M., S. Brand, B. E. Bouma, G. J. Tearney, C. C. Compton, and N. S. Nishioka. 2001. Diagnosis of specialized intestinal metaplasia by optical coherence tomography. *Gastroenterology* 120 (1): 7–12.

Sharma, P., A. Bansal, S. Mathur, S. Wani, R. Cherian, D. McGregor, A. Higbee, S. Hall, and A. Weston. 2006. The utility of a novel narrow band imaging endoscopy system in patients with Barrett's esophagus. *Gastrointestinal Endoscopy* 64 (2): 167–175.

Shim, M. G., and B. C. Wilson. 1997. Development of an *in vivo* Raman spectroscopic system for diagnostic applications. *Journal of Raman Spectroscopy* 28 (2–3): 131–142.

Shim, M. G., B. C. Wilson, E. Marple, and M. Wach. 1999. Study of fiber-optic probes for *in vivo* medical Raman spectroscopy. *Applied Spectroscopy* 53 (6): 619–627.

Standish, B. A., X. Jin, J. Smolen, A. Mariampillai, N. R. Munce, B. C. Wilson, I. A. Vitkin, and V. X. D. Yang. 2007. Interstitial Doppler optical coherence tomography monitors microvascular changes during photodynamic therapy in a Dunning prostate model under varying treatment conditions. *Journal of Biomedical Optics* 12 (3): 034022.

Suter, M. J., B. J. Vakoc, P. S. Yachimski, M. Shishkov, G. Y. Lauwers, M. Mino-Kenudson, B. E. Bouma, N. S. Nishioka, and G. J. Tearney. 2008. Comprehensive microscopy of the esophagus in human patients with optical frequency domain imaging. *Gastrointestinal Endoscopy* 68 (4): 745–753.

Tanaka, K., H. Toyoda, S. Kadowaki, R. Kosaka, T. Shiraishi, I. Imoto, H. Shiku, and Y. Adachi. 2006. Features of early gastric cancer and gastric adenoma by enhanced-magnification endoscopy. *Journal of Gastroenterology* 41 (4): 332–338.

Tang, S. J., M. L. Gordon, V. X. D. Yang, M. E. Faughnan, M. Cirocco, B. Qi, E. S. Yue, G. Gardiner, G. B. Haber, G. Kandel, P. Kortan, A. Vitkin, B. C. Wilson, and N. E. Marcon. 2003. *In vivo* Doppler optical coherence tomography of mucocutaneous telangiectases in hereditary hemorrhagic telangiectasia. *Gastrointestinal Endoscopy* 58 (4): 591–598.

Tearney, G. J., M. E. Brezinski, B. E. Bouma, S. A. Boppart, C. Pitris, J. F. Southern, and J. G. Fujimoto. 1997. *In vivo* endoscopic optical biopsy with optical coherence tomography. *Science* 276 (5321): 2037–2039.

Togashi, K., D. G. Hewett, D. A. Whitaker, G. E. Hume, L. Francis, and M. N. Appleyard. 2006. The use of acetic acid in magnification chromocolonoscopy for pit pattern analysis of small polyps. *Endoscopy(Stuttgart)* 38 (6): 613–616.

Uedo, N., R. Ishihara, H. Iishi, S. Yamamoto, T. Yamada, K. Imanaka, Y. Takeuchi, K. Higashino, S. Ishiguro, and M. Tatsuta. 2006. A new method of diagnosing gastric intestinal metaplasia: Narrow-band imaging with magnifying endoscopy. *Endoscopy* 38 (8): 819.

Vakoc, B. J., M. Shishko, S. H. Yun, W. Y. Oh, M. J. Suter, A. E. Desjardins, J. A. Evans, N. S. Nishioka, G. J. Tearney, and B. E. Bouma. 2007. Comprehensive esophageal microscopy by using optical frequency-domain imaging (with video){A figure is presented}. *Gastrointestinal Endoscopy* 65 (6): 898–905.

Wada, Y., S. E. Kudo, H. Kashida, N. Ikehara, H. Inoue, F. Yamamura, K. Ohtsuka, and S. Hamatani. 2009. Diagnosis of colorectal lesions with the magnifying narrow-band imaging system. *Gastrointestinal Endoscopy* 70 (3): 522–531.

Wallace, M. B., L. T. Perelman, V. Backman, J. M. Crawford, M. Fitzmaurice, M. Seiler, K. Badizadegan, S. J. Shields, I. Itzkan, R. R. Dasari, J. Van Dam, and M. S. Feld. 2000. Endoscopic detection of dysplasia in patients with Barrett's esophagus using light-scattering spectroscopy. *Gastroenterology* 119 (3): 677–682.

Wallace, M. B. 2009. Advances in imaging and technology of pre-invasive neoplasia: The big (and small) picture. *Gastroenterology* 137 (5): 1582–1583.

Wang, T. D., and J. Van Dam. 2004. Optical biopsy: A new frontier in endoscopic detection and diagnosis. *Clinical Gastroenterology and Hepatology* 2 (9): 744–753.

Watson, A. J. M., S. Chu, L. Sieck, O. Gerasimenko, T. Bullen, F. Campbell, M. McKenna, T. Rose, and M. H. Montrose. 2005. Epithelial barrier function *in vivo* is sustained despite gaps in epithelial layers. *Gastroenterology* 129 (3): 902–912.

White, B. R., M. C. Pierce, N. Nassif, B. Cense, B. H. Park, G. J. Tearney, B. E. Bouma, T. C. Chen, and J. F. De Boer. 2003. *In vivo* dynamic human retinal blood flow imaging using ultra-high-speed spectral domain optical doppler tomography. *Optics Express* 11 (25): 3490–3497.

Wong Kee Song, L. M., D. G. Adler, B. Chand, J. D. Conway, J. M. B. Croffie, J. A. DiSario, D. S. Mishkin, R. J. Shah, L. Somogyi, and W. M. Tierney. 2007. Chromoendoscopy. *Gastrointestinal Endoscopy* 66 (4): 639–649.

Yang, V. X. D., S. J. Tang, M. L. Gordon, B. Qi, G. Gardiner, M. Cirocco, P. Kortan, G. B. Haber, G. Kandel, I. A. Vitkin, B. C. Wilson, and N. E. Marcon. 2005. Endoscopic Doppler optical coherence tomography in the human GI tract: Initial experience. *Gastrointestinal Endoscopy* 61 (7): 879–890.

Yang, X., G. Jenkins, J. Franzke, and A. Manz. 2005. Shear-driven pumping and Fourier transform detection for on chip circular chromatography applications. *Lab on a Chip* 5 (7): 764–771.

Yao, K., A. Iwashita, H. Tanabe, T. Nagahama, T. Matsui, T. Ueki, S. Sou, Y. Kikuchi, and M. Yorioka. 2007. Novel zoom endoscopy technique for diagnosis of small flat gastric cancer: A prospective, blind study. *Clinical Gastroenterology and Hepatology* 5 (7): 869–878.

Zambelli, A., V. Villanacci, E. Buscarini, G. Bassotti, and L. Albarello. 2008. Collagenous colitis: A case series with confocal laser microscopy and histology correlation. *Endoscopy* 40 (7): 606–618.

Zanati, S., N. E. Marcon, M. Cirocco, N. Bassett, C. Streutker, G. P. Kandel, P. P. Kortan, S. Rychel, A. Douplik, and B. C. Wilson. 2005. Onco-life fluorescence imaging during colonoscopy assists in the differentation of adenomatous and hyperplastic polyps and improves detection rate of dysplastic lesions in the colon. *Gastroenterology* 128 (4 Suppl. 2).

Index

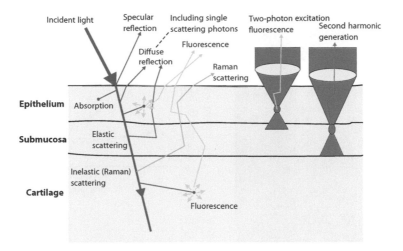

FIGURE 1.2 Schematic diagram of light pathways in bronchial tissue. A beam of incident light could interact with the bronchial tissue and generate various secondary photons measurable at the tissue surface: specular reflection, diffuse reflection, fluorescence, Raman scattering, two-photon excitation fluorescence, and second harmonic generation. These measurable optical properties can be used for determining the structural features as well as the biochemical composition and functional changes in normal and abnormal bronchial tissues. (Reproduced with permission from Zeng, H. et al. 2004. *Photodiagnosis and Photodynamic Therapy* 1: 111–122.)

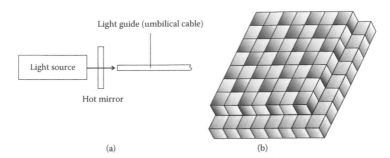

FIGURE 2.6 Principle of Bayer filtering color imaging. (a) simplified white light illumination; (b) Bayer filter mosaic on top of a CCD array.

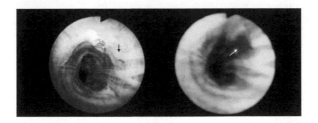

FIGURE 5.1 White light image (left) and autofluorescence image (right) of a carcinoma *in situ* lesion in the left main bronchus. The white light image shows a subtle nodular lesion. The fluorescence image shows a brownish area highlighted the lesion much better than the white light image. (Reproduced with permission from Zeng, H. et al. 2004b. *Photodiagn. Photodyn. Ther.* 1: 111–122.)

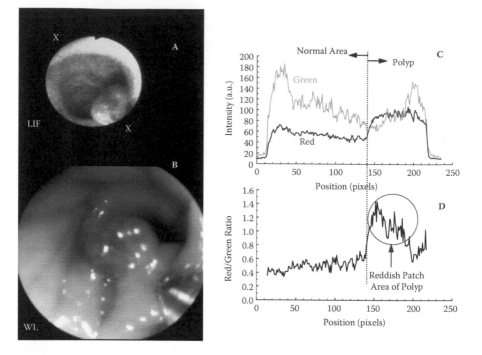

FIGURE 5.4 LIF image (A) and white light image (B) of a tubular adenoma polyp in the sigmoid colon. Part of the polyp appears reddish in the LIF image. Profiles of fluorescence intensities (C) and Red/Green ratio (D) along a line across the LIF image (A). The line starts at the "x" mark on the up left corner and ends at the "x" mark on the bottom right corner. (Reproduced with permission from Zeng, H. et al. 1998. *Bioimaging* 6: 151–165.)

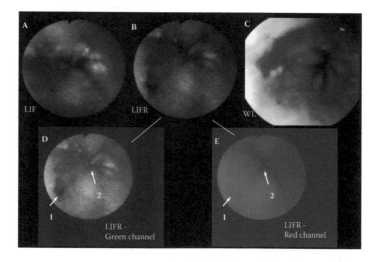

FIGURE 5.5 LIF image (A), LIFR image (B), and WL image (C) of an esophagus dysplasia. The separate green channel image (D) and red channel image (E) of the LIFR image are also shown. (Reproduced with permission from Zeng, H. et al. 1998. *Bioimaging* 6: 151–165.)

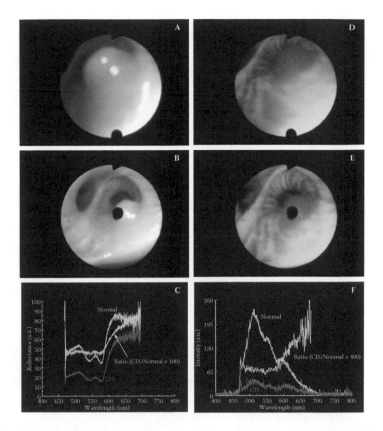

FIGURE 6.7 Example WL images, reflectance spectral curves (a,b,c) and FL images and spectral curves (d,e,f) of a *carcinoma in situ* lesion and its surrounding normal tissue. (Reproduced with permission from Zeng, H. et al. 2004. *Opt. Lett.* 29: 587–589.)

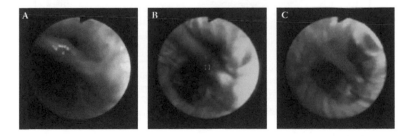

FIGURE 7.9 Pictures showing images acquired by a multimode bronchoscopy system. (A) white light image, and (B) blue light excited fluorescence image (Xillix, Onco-LIFE) of the same location as (A). For image (B) green is representative of normal tissue, and dark red (centre of image) is diseased tissue. (C) same lesion being excited simultaneously with blue light to generate a fluorescence image and with 785 nm light from the Raman catheter, which can be seen in the top right corner of the image. (Reproduced with permission from Short, M. et al. H. 2011. *J. Thoracic Oncol.* 6: 1206–1214.)

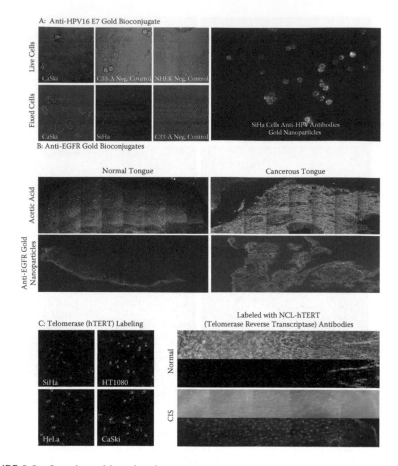

FIGURE 8.2 Imaging with molecular contrast.

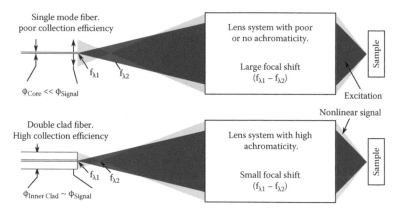

FIGURE 10.1 Fiber and lens considerations for nonlinear endomicroscopy. (a) Single mode fiber diameter (Φ_{Core}) is much smaller than the diameter of the cone of nonlinear optical signal (Φ_{Signal}) leading to suboptimal collection efficiency. Poor lens achromaticity leads to a large focal shift ($f_{\lambda 1} - f_{\lambda 2}$) between excitation and emission, exacerbating the problem. (b) Large inner cladding diameter of double clad fiber ($\Phi_{Inner\ Clad}$) and lens achromaticity both mitigate the problem, improving collection efficiency.

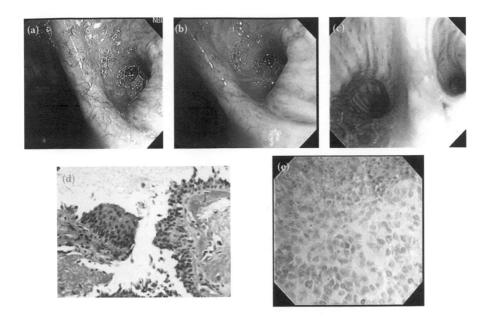

FIGURE 12.1 Evaluation of area of moderate dysplasia at RB6 with various optical imaging modalities. (a) High resolution and narrow band imaging showing complex networks of tortuous vessels, (b) high-resolution white light bronchoscopy, (c) autofluorescence imaging using AFI device; lesion appears magenta on green background, (d) histopathology of biopsy specimen showing moderate dysplasia, (e) endocystoscopic image of lesion. (Courtesy of Dr. Kiyoshi Shibuya, Matsudo City Hospital, Japan.).

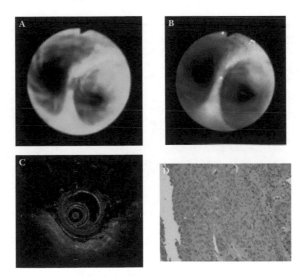

FIGURE 12.2 Area of carcinoma *in situ* at RB3 in 76yo male with COPD. (A) Area of CIS under autofluorescence imaging with OncoLife system (Xillix, Richmond, BC, Canada), (B) Area of CIS with standard white light bronchoscopy, (C) OCT image of area of CIS showing thickened epithelium with intact basement membrane, (D) Histopathology showing carcinoma *in situ*. (Courtesy of Professor Stephen Lam and Dr. Keishi Ohtani, BC Cancer Agency, Vancouver, Canada.).

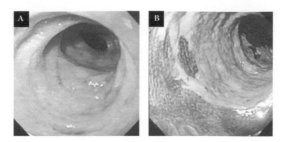

FIGURE 13.1 Lugol's solution used to detect squamous cell carcinoma. A: White light endoscopic view of esophagus. B: Endoscopic view following staining with Lugol's solution showing squamous cell carcinoma in the area that is unstained (yellow). The area stained dark greenish-brown is normal squamous mucosa.

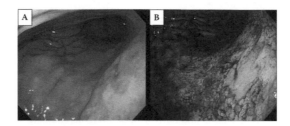

FIGURE 13.2 A: Conventional white light endoscopy of the stomach showing an irregular mucosal pattern with mild erythema. B: Chromoendoscopy with acetic acid indigocarmine mixture (AIM) showing the border of the lesion clearly and biopsy from this area showed tubular adenocarcinoma. (Courtesy of Koichi Okamoto, Tokushima University Hospital, Japan.)

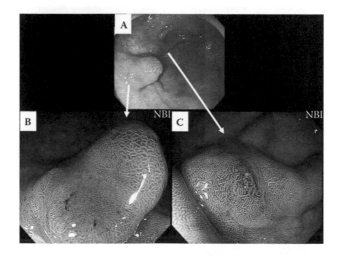

FIGURE 13.3 A: A couple of polypoid lesions were identified in the body of the stomach with WLE. B: NBI with magnification image showing smaller mucosal pattern in the central depressed area comparing to the surrounding area and the depressed area has abundant microvessels. Biopsy from the central area showed differentiated tubular adenocarcinoma. C: NBI with magnification image showing regular oval mucosal pattern and biopsy showed no malignant cell, which proved this as a scar due to the previous endoscopic resection. (Courtesy of Atsuo Oshio, National Hospital Organization, Kochi Hospital, Japan.)

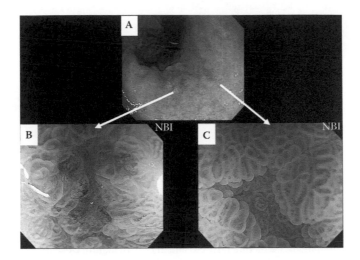

FIGURE 13.4 A: Reddish depressed lesions were identified in the body of the stomach with white light endoscopy. B: NBI with magnification image showing irregular microvascular pattern and biopsy showed differentiated tubular adenocarcinoma. C: NBI with magnification image showing regular microvascular pattern and biopsy showed focal chronic gastritis. (Courtesy of Atsuo Oshio, National Hospital Organization, Kochi Hospital, Japan.)

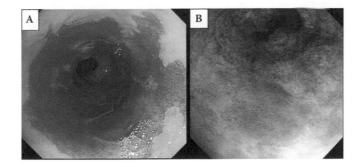

FIGURE 13.5 A: High resolution WLE image of the long segment of Barrett's esophagus. B: AFI image showing a patch of positive fluorescence (violet) at 6 o'clock. Biopsy from this area showed high-grade dysplasia. (Courtesy of L.M. Wong Kee Song, Mayo Clinic, Rochester, Minnesota.)

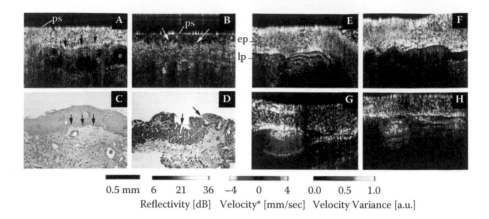

0.5 mm   6   21   36   −4   0   4   0.0   0.5   1.0

Reflectivity [dB]   Velocity* [mm/sec]   Velocity Variance [a.u.]

FIGURE 13.8   EDOCT images of Barrett's esophagus and esophageal varices from four different patients. (A,C) Subsquamous BE with mucosal glands (asterisks) underneath the clearly delineated epithelial-lamina propria interface (arrows). (B,D) Barrett's esophagus with superficial glandular structure (arrows) and microvasculature close to the surface. (E,F) Doppler images of blood flow in the dilated variceal vessels. The blood flow velocity in (E) is more than three times greater than that seen in normal esophagus. (G,H) Velocity-variance images of microcirculation in dilated variceal vessels. (C,D) CD34, Orig. mag. ×10. PS, External surface of the imaging tip. (Reprinted from Yang, V. X. D., Tang, M. L. et al.. 2005. *Gastrointestinal Endoscopy* 61 (7): 879–890. With permission from Elsevier.)

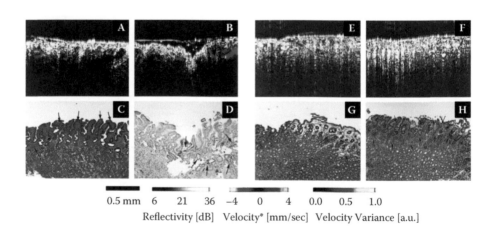

0.5 mm   6   21   36   −4   0   4   0.0   0.5   1.0

Reflectivity [dB]   Velocity* [mm/sec]   Velocity Variance [a.u.]

FIGURE 13.9   (A,B) Color EDOCT images of GAVE in two patients. (C) Consistent with the H&E staining, dilated microvasculature (arrows) is present immediately beneath the tissue surface. (D) Consistent with the CD34 staining, dilated microvasculature (arrows) is present immediately beneath the tissue surface. (E) Color EDOCT images of portal hypertensive gastropathy showing superficial vessels, consistent with the H&E histology in (G) and (H) marked by arrows. (F) Velocity variance EDOCT images of PHG, showing superficial vessels, consistent with the H&E histology in (G) and (H) marked by arrows. (C,D,G,H) Orig. mag. ×10. (Reprinted from Yang, V. X. D., Tang, M. L. et al., 2005. *Gastrointestinal Endoscopy* 61 (7): 879–890. With permission from Elsevier.)

T - #0405 - 071024 - C8 - 234/156/12 - PB - 9780367379070 - Gloss Lamination